HOLOGRAPHY - DIFFERENT FIELDS OF APPLICATION

Edited by **Freddy Alberto Monroy Ramírez**

Holography - Different Fields of Application
Edited by Freddy Alberto Monroy Ramírez

Published by InTech
Janeza Trdine 9, 51000 Rijeka, Croatia

Publishing Process Manager Dragana Manestar

Technical Editor Teodora Smiljanic

Cover Designer InTech Design Team

Additional hard copies can be obtained from orders@intechopen.com

Holography - Different Fields of Application, Edited by Freddy Alberto Monroy Ramírez
p. cm.
ISBN 978-953-307-635-5

Meet the editor

Dr. Freddy Alberto Monroy Ramírez was Born in Bogotá, Colombia. He is currently a professor at the Department of Physics of the National University of Colombia. He obtained his Doctorate in Sciences-Physics at the same University in the year 2008, with a dissertation entitled "Digital Holographic Microscopy with Phase Objects", in which he presented an experimental strategy based on holography and tomography for the decoupling of the morphological information and the refraction index contained in a map of different phases, obtained through digital holographic microscopy. In the areas of interferometry and digital holography he has developed various research projects; and currently, his research is dedicated to the development of a tool for palynological characterization through digital holographic microscopy. He has published various national and international publications in the areas of digital holography, digital holographic interferometry, and digital holographic microscopy. He has been a Colciencias referee for many research projects and has directed many Sciences-Physics Master theses in digital holography.

Contents

Preface

The word "holography" comes from the Greek word "*holos*" which stands for everything and "*graphos*" which means graphic, and this term has been coined for the registration and recuperation of the complex optical field that provides information about the surface of the object that reflects the light, or about the interior of the object through which light passes. Adding a reference beam to the light coming from the object that is being studied (by reflection or transmission) and then registering a pattern of interference between them, provides information from the point-to-point phase differences of the entire area of the image to be obtained, and these phase differences give the notion of three-dimensionality at the moment of reconstructing the hologram; this notion of three-dimensionality is what marks the difference between photography and holography. Recuperating the intensity as well as the phase of the complex optical field in holography, instead of only the intensity of the registered field as in a photograph, is the basic difference between these two recuperation processes of the information about the field that is reflected in or passes through an object. As a consequence, a greater quantity of information can be extracted from a hologram than from a photograph, as information is obtained about the three-dimensionality and the internal structure of the study object. The phase differences found in the field transmitted by a translucent object provide information about the morphology as well as the internal variations of the refraction index, which is primarily applicable to a biological sample that allows description of tissues, cells, pollen grains, etc. In the same way, in the case of opaque objects, the field reflected by them permits the information about the micro-topography and morphology of objects to be acquired, at the macroscopic level as well as the microscopic level. For these reasons, the study of holography has led to applications in very diverse branches of knowledge and has reinforced investigative and technological areas such as microscopy, non-destructive testing, security, information storage, etc. Due to the wide possibility of applications that are unleashed by holography, infinite literature currently exists at the basic level as well as the specialized level that demonstrates the importance of research in holography and its applications.

In this book, some differences will be pointed out from the typical scientific and technological literature about the theoretical study of holography and its applications, and therefore different topics will be shown which are neither very commercial nor very well-known and which will provide a distinct vision, evident in chapters such as: Electron Holography of Magnetic Materials, Polarization Holographic Gratings Formed on Polymer Dispersed Liquid Crystals, and Digital Holography: Computer-Generated Holograms and Diffractive Optics in Scalar Diffraction Domain.

The readers of this book will acquire a different vision of both the application areas of holography and the wide range of possible directions in which to guide investigations in the different optic fields.

Dr. Freddy Alberto Monroy Ramírez
Physics Department, Faculty of Science
The National University of Colombia, Bogota
Colombia

1

Holography – What is It About?

Dagmar Senderakova
Comenius University in Bratislava, Slovakia

1. Introduction

The 21st century is said to be a photon-century. People meet contemporary optics (holography, as well) applications everywhere. It would be appropriate to increase the common education level in this field for people to be able to understand new surrounding technologies, entering our everyday lives. Optics serves as an important part of many scientific experimental methods. This way, such information could be useful also for researchers without a professional optical education.

Let us follow, briefly, at least, the history of a "mystery of light". Light surrounds people from the very beginning. Man lives in the world bathing in light. The eye is said to bring us the greatest piece of information. Naturally, people have been taken an interest in that "something", useful for our eyes, called light. Thousands of years after human started with using fire to illuminate nights (~ 12000 BC), Indians, Greek and Arab scholars began to formulate theories on light (Davidson, 1995). Man had been taken a great interest in "optical experiments". Even in 423 BC, Aristophanes wrote a comedy, Clouds, in which an object was used to reflect and concentrate the sun's rays and to melt an IOU recorded on a wax table (Stevenson, 1994).

For now, let us briefly mention only some steps, dealing with the original question: "*What is light?*" Such a question is closely related to the answer – "*How can man see*?" So-called "tactile theory" seemed to be the first one that gave an answer. It was based on the assumption that the human eye sent out invisible probes/light rays "to feel" objects (Plato, Euclid, 400-300 BC).

However, people cannot see in dark! Aristotle (350 BC) was among the first to reject such a theory of vision. He advocated for a theory by which the eye received rays rather than directed them outward. Such a theory, so-called "*emission theory*", appeared later and offered a solution of the paradox, mentioned above. It stated that bright objects sent out beam of particles into the eye.

More than 17 centuries passed, while experiments with light, mirrors and lenses had led to construction of microscopes and telescopes, which broaden the worldview of early scientists. As for the character of light, Ch. Huygens was able to explain many of the known propagation characteristics of light, using his *wave* theory. He assumed light to transmit through all-pervading ether that is made up of small elastic particles, each of which can act as a secondary source of wavelets. However, he could not explain such a simple thing, like rectilinear propagation of light...

The development culminated in 1704, when I. Newton published his Optiks and advocated his *corpuscular theory: Light is a system of tiny particles that are emitted in all directions from a*

source in straight lines. The corpuscles are able to excite waves in ether. Light slows when entering a dense medium. This theory is used to describe reflection. However, it cannot explain some atmospheric phenomena, like supernumerary bow, the corona, or an iridescent cloud.
In about 100 years later, the Newton's corpuscular theory was overturned by *the wave theory* when demonstrating and explaining phenomena of interference, diffraction and polarisation of light (T. Young, A. J. Fresnel, D. F. Arago, J. Fraunhofer).
Then a new era began. The discovery of electromagnetic waves is considered perhaps, the greatest theoretical achievement of physics in the 19th century. Besides, the speed of light had been known to be about 300 000 km/s, since the 17th century (Olaus Roemer, a Danish astronomer).
James Clerk Maxwell (1831-79) completed his formulation of the field equations of electromagnetism to be applicable also for space without wires. Moreover, he calculated that the speed of propagation of an electromagnetic field is approximately that of the speed of light. Because of that he proposed the phenomenon of light to be an electromagnetic phenomenon. This way, Maxwell established the theoretical understanding of light.
In the late 19th century it was believed that all the electromagnetic phenomena could be explained by means of this theory. However, an unexpected problem arose when one tried to understand the radiation from glowing matter like the sun, for example. The spectral distribution did not agree with the theories based on Maxwell's work. There should be much more violet and ultraviolet radiation from the sun than had actually been observed.
Max Planck came with a solution. Being skilled in mathematics, he played around with the equations and introduced mathematically, only, the idea of energy quantum $h\nu$, where $h = 6.626 \times 10^{-34}$ Js is now called Planck's constant. He assumed that for a wave with a certain frequency ν, it was only possible to have energies that were multiples of $h\nu$. Such a math trick caused the new calculation to agree with experiment perfectly but nobody really believed that light came in "particles".
Time went and showed that energy could only come in small "parcels" of $h\nu$. These small parcels of light were called *quanta*. Einstein liked the idea of quanta and supported their existence explaining *photoelectric effect* and describing light-matter interaction via *absorption and spontaneous and stimulated emission,* which initiated birth of a new kind of a light source – *laser* (*Light Amplification by Stimulated Emission of Radiation*).
Einstein extended the *quantum theory* of thermal radiation proposed by Max Planck to cover not only vibrations of the source of radiation but also vibrations of the radiation itself.
As for photon – let's mention something interesting. Einstein did not introduce the word photon. It originated from Gilbert N. Lewis, years after Einstein's works on photoelectric effect. He wrote a letter to the Nature magazine editor (Levis, 1926): "*...I therefore take the liberty of proposing for this hypothetical new atom, which is not light but plays an essential part in every process of radiation, the name photon....*" Interestingly, Lewis did not consider photons as light or radiant energy but as the carriers of radiant energy.
The quantum theory also met with difficulties. Solving them, quantum electrodynamics (QED), was developed (Nobel Prize in Physics 1965 to S. Tomonaga, J. Schwinger and R. P. Feynman). It became the most precise theory in physics and contributed especially to development of particle physics. However, in the beginning it was not judged necessary to apply QED to visible light.
Later, it was just the development of lasers, sources of *coherent* light and similar devices, which caused a more realistic description to be required when considering the light from a

thermal source (light bulb, sun, and so on) comparing with that of laser. Their light waves seemed to be much more chaotic and it seemed easier to describe the disorder that stemmed from it as randomly distributed photons.
One half of the 2005-year's Nobel Prize in Physics was awarded to Roy J. Glauber for his pioneering work in applying quantum physics to optical phenomena. He had developed a method for using electromagnetic quantization to understand optical observations. He carried out a consistent description of photoelectric detection with the aid of quantum field theory, which laid the foundations for the new field *of Quantum Optics*. It soon became evident that technical developments made it necessary to use the new quantum description of the phenomena, too. Man can get completely new technical applications of quantum phenomena, for example to enable safe encryption of messages within communication technology and information processing.
Let us remember that light still is the same light, despite speaking about particles, waves, quanta, and so on. In physics, we are trying to describe and explain our observations. Moreover, after having understood the observed phenomenon, we would like to predict another phenomenon and prove it experimentally. To do that, we have to use a "language". It is math. We can use math to describe behaviour of material objects, waves and various kinds of energy. Taking into account the experience, man uses the known "items" to create a model of the observing object - light in our case. Just that is the origin of all the wave models of light, mentioned above. There are optical phenomena (e. g. reflection and refraction of light) explained easily to compare them to the behaviour of mechanical corpuscles or rays. To explain phenomena of interference, diffraction and polarisation of light, the model of waves has to be considered. P*hotons*, seem to be very useful to explain absorption and scattering of light, photoelectric effect in a simply way, and so on.

2. Wave aspects of light

Taking into account the main goal of the chapter - to understand the new and attention catching problem of holographic recording, let us get more familiar with the wave model of light. Namely, making a hologram means to deal with interference of light. To see, what is the hologram about, we need diffraction of light, in fact. Both the phenomena are usually described using the math language for waves.

2.1 Wave

In physics, a wave is defined to be a process of disturbance travelling throughout a medium. How is it performed, it depends on the kind of disturbance and on the medium-disturbance coupling. Wave transfers energy from one particle of medium to another one without causing a permanent displacement of the medium itself.
Let us have a look at a light wave, being modelled by an electromagnetic wave. It is enough to deal with the electric wave. The magnetic one is related to it by Maxwell's equations. Conventionally, amplitude of electric field vector f can be expressed in the form

$$f = A.\cos\phi \qquad (1)$$

The peak value A of the alternating quantity f is called *amplitude*. The sign ϕ denotes phase of the wave. It determines development of the periodic wave. Let us have a wave propagating in direction z. It varies in both, space (z) and time (t), which are included just in the phase

$$\phi = \omega t - kz + \phi_0 \tag{2}$$

ϕ_0 is the initial phase of the wave at $z = 0$ and $t = 0$. $k = 2\pi/\lambda$ defines *wave number*. It is the absolute value of *wave vector* **k** determining direction of wave propagation. The distance between the two neighbouring amplitude peaks of the same kind is *wavelength* λ [m]. ω [rad.s^{-1}], is *angular frequency*, which is related to the linear *frequency*, ν [s^{-1}], by the formula $\omega = 2\pi\nu$. It lasts $T = \lambda/c$ seconds to pass the path λ at the speed c. Such a time interval is called *period* and $T = 1/\nu$.

A *wave front* is another useful term for us, else. It is the surface upon which the wave has equal phase. It usually represents the peak amplitude of the wave and is perpendicular to the direction of propagation, i.e. to the wave vector **k**. Wave fronts related to the same phase are separated by the wavelength.

Considering a wave propagating in the +z direction, the wave vector **k** is parallel to the z-axis everywhere. Because of that, wave fronts are parallel planes, perpendicular to the z-axis. Such a wave is known as a *plane wave*. When the $\boldsymbol{k}(k_x, k_y, k_z)$ direction is general, the phase (2) in a point determined by a displacement vector $\boldsymbol{r}(x, y, z)$ includes the scalar product of $\boldsymbol{k}.\boldsymbol{r}$ instead of kz.

Let us mention also a *spherical wave*

$$f(r,t) = \frac{A}{r}\cos(wt - kr + \phi_0) \tag{3}$$

which is irradiated from a point light source in homogeneous medium. In such a case wave fronts are centrally symmetrical spheres, so it is enough to consider only radial coordinate r of *spherical* ones. Moreover, **k** and **r** are parallel, so $\boldsymbol{k}.\boldsymbol{r} = kr$. Increasing distance from the source the surface of the sphere increases and amplitude A decreases proportionally to $1/r$.

We are going to work with waves in this topic and it has been known that using trigonometric functions leads to cumbersome calculations. To overcome such a problem, a complex notation is used. The trigonometric function can be replaced by exponential functions applying Euler's formula

$$e^{i\phi} = \cos\phi + i\sin\phi \qquad e^{-i\phi} = \cos\phi - i\sin\phi = (e^{i\phi})^* \tag{4}$$

Such a way will simplify the mathematical description of light greatly. $e^{i\phi}$ is a complex function, $\cos\phi$ is its real part, $\sin\phi$ is its imaginary part and i is imaginary unit. $(e^{i\phi})^*$ is said to be a complex conjugate function to $e^{i\phi}$. From such a point of view the expression (1) can be considered as a real part of the complex function

$$f = Ae^{i\phi} = A\cos\phi + iA\sin\phi \tag{5}$$

When comparing to mechanics and electricity, there is a special property of light waves. The instantaneous amplitude f, which varies with both, time and space, cannot be measured experimentally in a direct way. The frequency of the *light wave* is too high for any known physical mechanism (photo electrical effect) to reply to the changes of the instantaneous amplitude f.

Any known detector replies only to the incident energy. When denoting energy transferred by a wave as w, it can be got as square of the amplitude, e.g. $w = A^2 = f.f^*$, when using the complex representation. The value known as *intensity* I of light, is proportional to the energy per unit of surface and unit of time. It is very important to realise that the time averaged

light intensity is a measurable value, only. Because of that it is said that both, light detection and light recording are *quadratic*.
Before starting with the basic phenomena of *interference* and *diffraction* of light, remember the simple wave model, describing the propagation of light wave through a space, else. The Dutch physicist Christian Huygens formulated a principle. It says that *each point on the leading wave front may be regarded as a secondary source of spherical waves, which themselves progress with the speed of light in the medium and whose envelope constitutes the new wave front later. The new wave front is tangent to each wavelet at a single point.*

2.2 Interference of light

Let us add two waves, i.e. illuminate a surface by two light beams. The observable result depends on what light beams were used. Mostly, one can observe a brighter surface comparing to that illuminated by one-beam, only. However, there are situations, when one can see both, parts of the surface with very high brightness, and parts with very low one, even dark. Just that case, when a kind of redistribution of all the incident energy can be observed, represents what is said to be the *interference of light*. Let us find what is the reason of such a redistribution of light energy when overlapping two light beams.
In the beginning, let us consider two light waves, f_1 and f_2, expressed by (5)

$$f_1 = A_1 exp(i\phi_1) \text{ and } f_2 = A_2 \exp(i\phi_2) \qquad (6)$$

They can meet at a time at every point of the surface with a phase difference $\Delta\phi = \phi_2 - \phi_1$. Let us find what can be observed. Taking into account quadratic detection of light, the result can be expressed by

$$< I > \sim < (f_1 + f_2)(f_1 + f_2)^* > = < |A_1|^2 + |A_2|^2 + 2A_1 A_2 \cos(\phi_2 - \phi_1) > \qquad (7)$$

Brackets < > represent time averaged light intensity I, $|A_i|^2 = I_i$ denote intensities of each of two waves.
Let us have a more detailed look at the phase difference $\Delta\phi = \phi_2 - \phi_1$. Taking into account the relation (2), the phase difference can be expressed in the form

$$\Delta\phi = \phi_2 - \phi_1 = (\omega_2 t - k_2 z_2 + \phi_{02}) - (\omega_1 t - k_1 z_1 + \phi_{01}) \qquad (8)$$

It is obvious that the time independence of the phase difference in (7) is the crucial condition to get interference of light, i.e. to observe and record energy distribution following the phase difference at any point of the surface. The conditions, being necessary to be fulfilled, follow from (8): $\omega_1 = \omega_2$, i.e. $\lambda_1 = \lambda_2$ and $\phi_{01} = \phi_{02}$. Such two waves are said to be *coherent* and only in such a case the intensity distribution (7) can be observed and recorded. Both the waves must have the same properties, i.e. the same wavelength, the same initial phases. Both waves have to come from one coherent light source. It is *laser*, where *stimulated emission* (Smith et al., 2007) takes part.
Relation (7) can be used to find the well-known conditions when either maximum or minimum of average intensity occurs:

$$< I > = I_{\max} \quad when \quad \Delta\phi = 2m\pi \quad i.e. \quad n.\Delta l = m\lambda, \quad m = 0,1,2,... \qquad (9a)$$

$$< I > = I_{\min} \quad when \quad \Delta\phi = (2m+1)\frac{\pi}{2} \quad i.e. \quad n.\Delta l = (2m+1)\frac{\lambda}{2} \quad m = 0,1,2,... \qquad (9b)$$

Product of index of refraction n and path difference Δl, which can be found in the phase difference, is known as *optical path difference*. Namely, $\Delta\phi = (2\pi/\lambda)n\Delta l$, when $\omega_1 = \omega_2$, and $\phi_{01} = \phi_{02}$,. The wavelength λ is taken in vacuum.
On the contrary, when the phase difference between two being added light waves is time dependent, the last term in (7) turns into zero. The average intensity distribution does not depend on the phase difference and no intensity distribution is observed, no interference occurs. Such two waves are said to be *incoherent*. However, real waves are *partially coherent*.
Concluding this part, let us give some notices dealing with *coherence*. Generally, it is defined by the correlation properties between quantities of an optical field. Interference is the simplest phenomenon revealing correlations between light waves. A *complex degree of mutual coherence* $\gamma_{12}(\tau)$ is defined to express the coherence of an optical field (Smith et al., 2007). Numbers 1 and 2 denote two point sources of interfering waves, and τ represents their relative delay. It can be shown that the interference pattern (7) is influenced by *module of degree of mutual coherence* $|\gamma_{12}(\tau)|$

$$< I >\sim I_1 + I_2 + 2\sqrt{I_1 I_2}\,|\gamma_{12}(\tau)|\cos(\phi_2 - \phi_1) \qquad (10)$$

In another words, *visibility* $V = (I_{max} - I_{min})/(I_{max} + I_{min})$ of interference pattern, which can be measured experimentally (Fig. 1), tells us about the *module of degree of mutual coherence* $|\gamma_{12}(\tau)|$

$$V_{12}(\tau) = \frac{I_{max} - I_{min}}{I_{max} + I_{min}} = 2|\gamma_{12}(\tau)|\left[\sqrt{\frac{I_1}{I_2}} + \sqrt{\frac{I_2}{I_1}}\right]^{-1} \qquad (11)$$

A practical measurement of the degree of coherence amounts to creating an interference pattern between two waves (1, 2), or of a wave with itself. Temporal $V_{11}(\tau)$ and spatial $V_{12}(0)$ dependencies can be obtained experimentally by varying either the delay τ using a moving mirror in an interferometer or keeping $\tau = 0$ and varying the distance between the point sources 1 and 2.

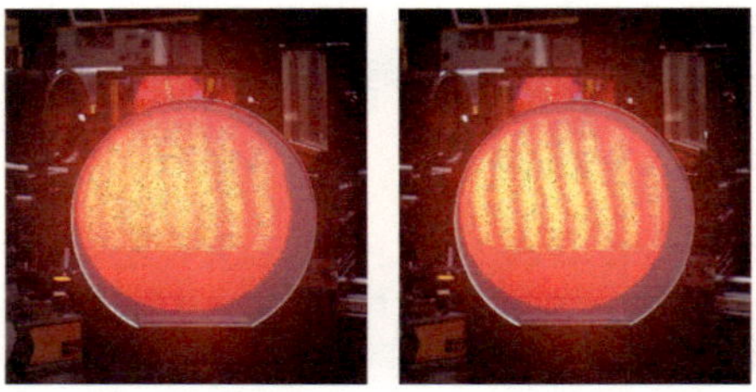

Fig. 1. Various visibility of interference pattern

This way either normalised Fourier transform of the frequency spectrum irradiated (related to the *temporal coherence*) or normalised Fourier transform of angular intensity distribution (related to the *spatial coherence*) can be obtained experimentally.
In the case of equal average intensities of both the waves, the module of the complex degree of coherence is given directly by the visibility V of the interference pattern.

2.3 Diffraction of light

Diffraction has been known as another phenomena of wave optics. Any deviation from rectilinear propagation of light that cannot be explained because of reflection or refraction is included into diffraction. When light passes through a narrow slit, it seems as it "bends" and incidents on the screen behind the slit also where darkness was expected to be according a geometric construction. Moreover, it is not a continuous illumination. Some fringes can be seen.

In fact, it is a result of interference of light, again. Let us have a look at what happens. For simplicity, a coherent plane wave, incident perpendicularly at the plane of the slit, passes the slit (Fig. 2.). The part of wave front restricted by the slit contains infinite number of point light sources having the same phase. Imagine the sources as divided into two equal groups - the sources above and under the z-axis. Infinite number of couples S_1 and S_2 just distant $b/2$ (half of the slit width) from each other can be found in the slit. Let us consider only light propagating at the same angle α from both the sources. Interference should occur very far away, at $L = \infty$, and can be observed in the second focus plane of a lens.

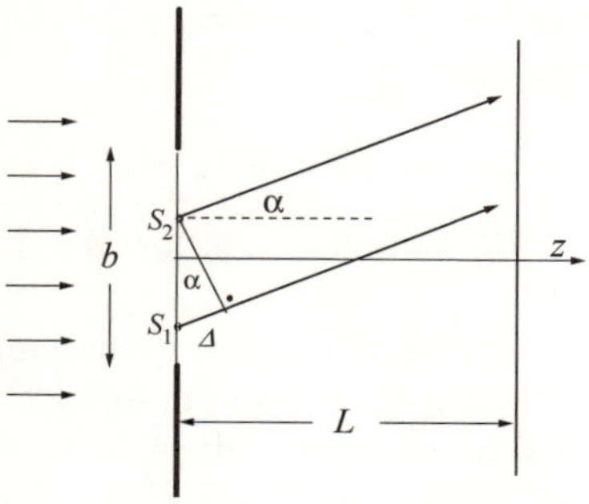

Fig. 2. Light passes through a slit (width of the slit is denoted by b)

Let us calculate the path difference Δl (Δ in Fig. 2) between interfering waves. It is given by the angle α and the slit width b

$$\Delta l = \frac{b}{2}\sin\alpha \tag{12}$$

Relations (9a) and (9b) determine angles α at which either interference maxima (constructive interference) or minima (destructive interference) occur.

The same analysis can be used for any such a pair of point light sources from the slit. It will increase the amount of energy propagating at the angle α.

The average relative intensity distribution (Fig. 3) relation

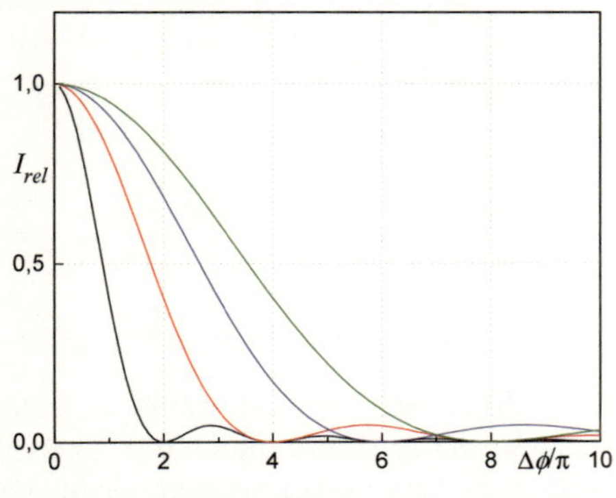

Fig. 3. Diffraction pattern at various slit widths ($\Delta\phi/\pi = (b/\lambda)\sin\alpha$, $b > b > b > b$)

$$\frac{I(\alpha,\lambda,b)}{I_0} = \left[\frac{\sin\left(k\frac{b}{2}\sin\alpha\right)}{k\frac{b}{2}\sin\alpha} \right]^2 \quad (13)$$

can be found analytically, applying the scalar diffraction theory (Smith et al., 2007). Practically usable are especially the conditions for interference minima

$$b.\sin\alpha = m\lambda,\ m = 1,2,3,... \quad (14)$$

Most of the light energy (~84%), which passed through the slit, is concentrated near the axis. It is called the zero order maximum. The apex angle 2α of this cone depends on the slit width and wavelengths of the used light

$$b.\sin\alpha = 1.\lambda \Rightarrow \sin\alpha = \lambda / b \quad (15)$$

It can be seen from the relation (15) – the less is b, the grater is $\sin\alpha$ and the greater is the angle α.

Practically, it is worth to notice that diffraction by a circular aperture is very important. Why is it? All the optical devices, like cameras, and so on, restrict the passing light by a circular aperture. The relation, similar to the (14) one, has the form (Smith et al., 2007)

$$\sin\alpha = 1.22\,(\lambda / D) \quad (16)$$

where D is the diameter of the aperture.

Let us notice another kind of diffraction, else, widely used, especially in spectroscopy (looking for various wavelengths) – *diffraction by a grating*. The grating is an ensemble of single equal slits, parallel to each other and having the same distance between each other. There are two parameters, which define the grating – the slit width (b) and the grating interval (d) – distance between centres of any two adjacent slits (Fig. 4.).

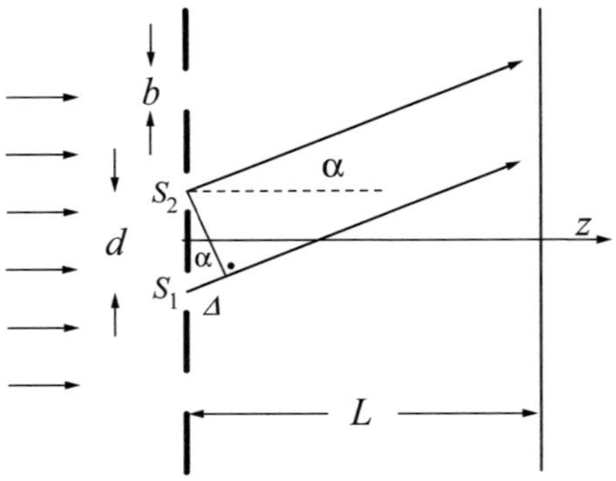

Fig. 4. Diffraction by a grating (d – grating interval, b – slit width)

It is the interference of many "diffractions" by single slits, in fact. The number of interfering "diffractions" depends on the number of illuminated slits. To explain the result, in a simply way we can use the analysis above when considering the single slit only. However, now any slit is considered to represent a single light source with the same phase.

Following the consideration above and the relation (12), the path difference between waves from two neighbouring sources (slits), propagating at an angle α can be expressed in the form

$$\Delta l = d.sin\alpha \tag{17}$$

Let us compare the relations for a single slit (12) and for a grating (17). Since d is defined to be the distance between two related points, e.g. the centres of two neighbouring slits, certainly $d > b/2$. Compare the angles for the single slit α_s and for the grating α_g, when destructive interference appears the first time, i.e. when $m = 1$ in (9b), we realize that

$$sin\alpha_s = \lambda / b \quad sin\alpha_g = \lambda / 2d \;\Rightarrow\; \alpha_s > \alpha_g \tag{18}$$

Relation (18) tells us that the destructive interference with the grating occurs at the less angle α_g than in case of one of grating slits (α_s). Because of that some intensity maxims can occur in the frame of the zero order maximum of a single slit (Fig. 5.). The number of intensity maxims $m = d/b$ can be found in an easy way (Smith et al., 2007) using relations (18).

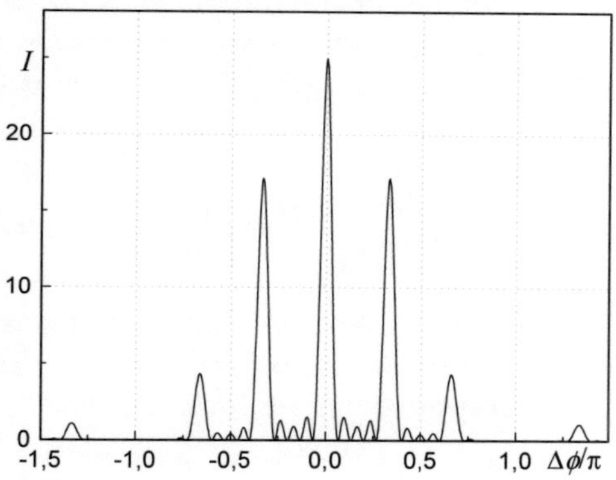

Fig. 5. Intensity distribution for $d/b = 3$, $\Delta\phi/\pi = (d/\lambda)\sin\alpha$

Let us also have a note to the influence of number of illuminated slits on the grating diffraction pattern (another name for the interference intensity distribution). It is related to many-beam interference (Smith et al., 2007). The more slits, the more beams and the highest and narrowest the intensity maximum. Of course, the values of all the intensity maxims are not equal. They are modulated by diffraction by the single slit.

The average relative intensity distribution relation can be found in the form (Smith et al., 2007)

$$\frac{I(\alpha,\lambda,b,d)}{I_0} = \left[\frac{\sin\left(k\frac{b}{2}\sin\alpha\right)}{k\frac{b}{2}\sin\alpha}\right]^2 \left[\frac{\sin\left(kN\frac{d}{2}\sin\alpha\right)}{\sin\left(k\frac{d}{2}\sin\alpha\right)}\right]^2 \tag{19}$$

Practically usable is especially the condition for interference maxims

$$d.\ sin\alpha_m = m\lambda,\ m = 0,1,2,3,\ ... \tag{20}$$

Concluding, let us mention the main principle of solving diffraction problems briefly, at least. Exact solutions are given by solving Maxwell's equations. However, well-known Kirchhoff's scalar theory gives very good results if period of diffraction structure does not approach a wavelengths size and amplitude vector does not leave a plane.

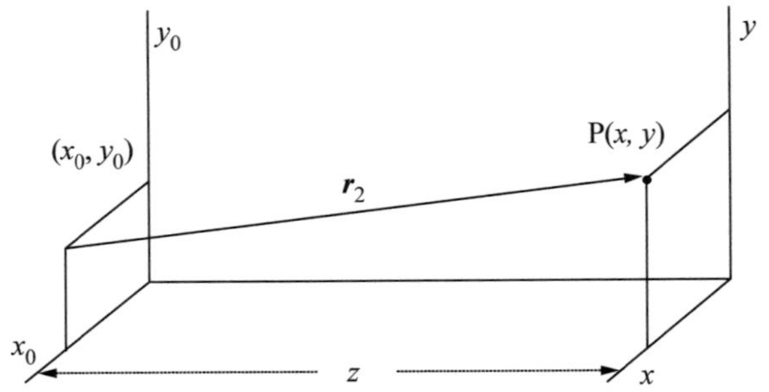

Fig. 6. To the principle to solve scalar diffraction problems

Let the plane (x_0, y_0) is the plane of the slit and diffraction is observed in the plane (x, y). To find resulting amplitude in $P(x, y)$, amplitudes of spherical waves from all the point sources in the plane of the slit have to be summed (Fig. 6). The idea is expressed by Kirchhoff's diffraction integral

$$f_P(x,y,z) \sim \iint_S \frac{f_P(x_0,y_0)}{r_2} \exp\{-ikr_2\} \, dx_0 \, dy_0 \tag{21}$$

where the distance $r_2 = [(x - x_0)^2 + (y - y_0)^2 + z^2]^{1/2}$ and S is size of the obstacle (slit). The experimentally observable interference pattern is given by $I(x, y, z) \sim f_P(x, y, z).[f_P(x, y, z)]^*$. To calculate the integral (21), two approximations for r_2 are used.

1. *Fresnel diffraction*

   $x - x_0 \ll z$, $y - y_0 \ll z$, i.e. next transformations are used
   $|\mathbf{r}_2| = [z^2 + (x - x_0)^2 + (y - y_0)^2]^{1/2} \to z + (x - x_0)^2/2z + (y - y_0)^2/2z$ — in the phase
   $|\mathbf{r}_2| = z$ — to express amplitude decreasing

In paraxial approximation (k_x, k_y << k_z) integral (21) turns into

$$f_P(x,y,z) = \frac{i}{\lambda z} \exp(-ikz) \iint_S f_0(x_0,y_0,0) \exp\left\{-\frac{ik}{2z}\left[(x-x_0)^2 + (y-y_0)^2\right]\right\} dx_0 \, dy_0 \tag{22}$$

2. *Fraunhofer diffraction*

   x_0, y_0, $\ll z$, i.e. next mathematical approximation is used to express the phase

$$(x-x_0)^2 \to x^2 - 2xx_0 \quad \text{and} \quad (y-y_0)^2 \to y^2 - 2yy_0$$

and integral (21) turns into

$$f_P(x,y,z) = \frac{i}{\lambda z} \exp\left\{-ik\left(z + \frac{x^2+y^2}{2z}\right)\right\} \iint_S f_0(x_0,y_0,0) \frac{\exp\{-ik(xx_0+yy_0)\}}{z} dx_0 \, dy_0 \tag{23}$$

Just that is the approximation useful while explaining the basics of holography. Real calculation of integral (23) gives relations (13) and (19).

3. Holography

After invention of coherent light sources - *lasers*, a new method, called *holography*, has been talking about. Basically, it is said to be a new method utilising light to record information. All of us have known a method using light to record information. It is photography. Both methods are kinds of optical recording. Why do we speak about holography? Is there something else comparing to photography?

All of you certainly enjoyed nice photos, marvellous pictures, which either remembered you of something pleasant or showed you something interesting, you have never seen before. Despite the plain shape of a photo, we are able to see and perceive a space on it. However, such ability of perceiving is only a consequence of our everyday experience of perspective. We are able to perceive depth of surrounding space since we have two eyes separated by a distance horizontally. Each eye sees an object in front of us from a bit different direction. The images created by the eye lenses differ a bit and thankful to sophisticated and still not completely understood "image processing" by our brain, we perceive a space. However, "3D impression" of photos can be exalted by stereo photography. In such a case we prepare for each of our eyes a special image, as it was in reality.

On the other side, when observing a hologram, one does not need to be experienced in anything. Simply, it is a 3D scene, indeed. Moving your head a bit allows you reveal even hidden objects when observing the hologram. What is the reason of such differences? To make a record, light was used each time.

The secret is encoded in the name *hologram*, in fact. The name was coined by British scientist Dennis Gabor (native of Hungary), who developed the theory of holography while dealing with the problem "how to improve resolution of an electron microscope". It comes from two Greek words *holos* (whole) and *gramma/graphe* (message/recording).

Such an origin gives a hint about recording "everything" of the light coming from the object. Another notice - do we not record light in whole when taking a picture by a photo-camera? What does it mean "in whole" and "not in whole"?

Let us remember a light wave and its properties (1). When expressing it, some attributes are included: *amplitude* and *phase*. Moreover, remember, again - the light wave cannot be recorded directly. Only the energy transferred by a wave, $w = |A|^2$, is a measurable and detectable value.

Just that is used when taking a classical picture. Optical system of a camera produces the image of every point of the object on the recording plane, where film/pixels are placed and influenced by incident light. That means - only information dealing with amplitude of the light wave was recorded and used to produce a record. The phase (2)

$$\phi = \omega t - kz + \phi_0$$

in which there is the variable z, telling us about the path of the light wave, i.e. from which distance the wave came, is lost. And just there is information about 3D properties of the object hidden. Light waves come to the recording medium with different phases (since passing different paths) from different points of the object.

However, a light wave cannot be recorded directly. In other words – the phase of the wave cannot be recorded directly, too. Only average intensity, proportional to energy transferred by the wave can be recorded. What is a solution?

3.1 Hologram recording

In 1947 Dennis Gabor found the solution. Since average light intensity can be recorded only, nothing about the phase when the light wave is alone can be recorded. On the other hand, when adding two coherent waves, the resulting intensity at any point depends on the phase difference between two waves at that point. We can record an intensity distribution – *interference pattern*, as mentioned in the part 2.

This way it is possible, as is shown later, to get from a hologram the same light wave as propagated from the 3D object, so the 3D object can be observed, indeed.

So the first phenomenon as the principle of the holography is *interference* of light waves. That demands coherent light waves. The simplest way how to get coherent light is – to use laser.

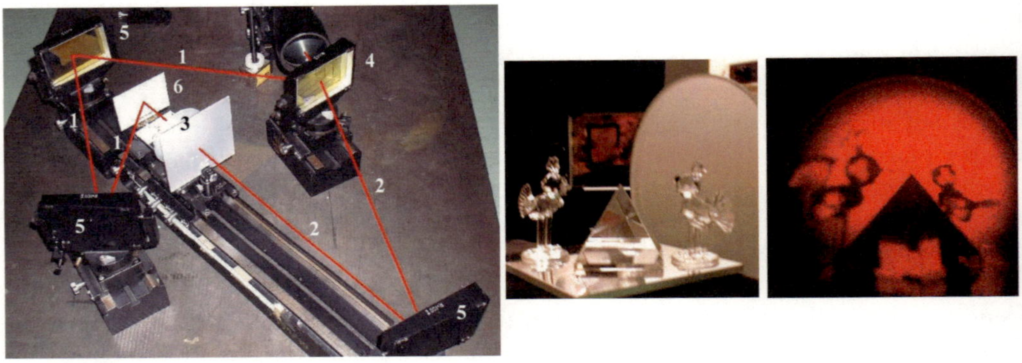

Fig. 7. Hologram recording set-up, object (a glass scene on a mirror), and reconstruction (1–reference wave, 2–wave to create the object wave, 3–object, 4–beam splitter, 5–mirrors, 6–recording medium)

Fig. 7 demonstrates the experimental set-up to record a hologram. Beam splitter 4 divides the laser beam into two parts. The wave 1 proceeds without any changes towards the recording medium 6. There is no information in this wave. It is the *reference* wave (***r***). Mostly, a plane wave is used, in our set-up too, but it is not a necessity. The wave 2 interacts with the object and *object wave* (***o***) is created. It is reflected or scattered by the object. It also can pass through the object. It depends on what kind of object we have. In our case the object is transparent, it is a glass scene on a mirror. Collimated laser beam scattered by ground glass and passing through the pyramid and birds represents the object wave.

Reference wave and object wave are directed by mirrors 5, meet each other and interfere in space, where the recording medium 6 is placed. The interference pattern is recorded. To get a hologram, the recording medium is exposed by incident beams and properly processed.

Besides coherent light, there are another experimental conditions else, which have to be fulfilled. All the set-up has to be stable. The path difference between interfering waves must not change even in $\lambda/2$. Such a value changes constructive interference into destructive one, intensity maximum changes into minimum and no interference pattern is recorded.

Moreover, a special holographic recording medium has to be used. The interference pattern of high density (of about 1000-3000 lines/mm) is recorded and the medium has to be able to record it. The density of the interference pattern depends on the angle between interfering waves and one can calculate it approximately using the relation (9a).
To find the two-beam interference pattern density along the *x*-direction, the k_x component of the wave vector $\boldsymbol{k}$ has to be considered (Fig. 8.) and (9a) gets the form

$$\Delta\phi = \frac{2\pi}{\lambda}(\sin\alpha_1 + \sin\alpha_2)x = 2m\pi$$

The difference Δx between two adjacent maxims can be expressed by the difference

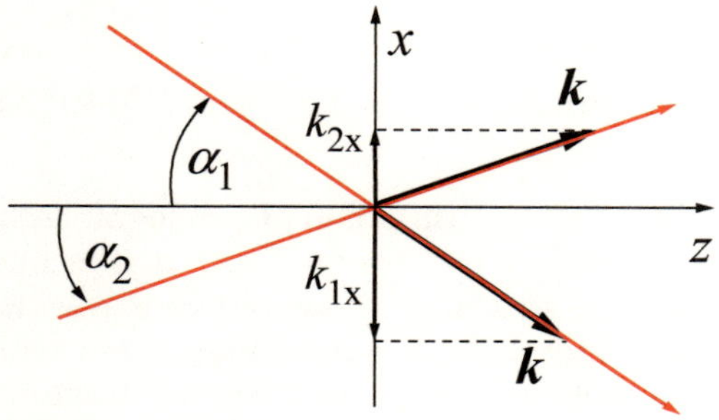

Fig. 8. To get interference pattern density

$$\Delta x = x_m - x_{m+1} = \frac{\lambda}{(\sin\alpha_1 + \sin\alpha_2)} \tag{24}$$

Reciprocal value of (24) gives the *spatial frequency*.
It would be useful to stop for a while at physical meaning of the process of recording. To record light a recording medium is used. It can be any medium, optical properties of which vary with the intensity of incident light. The intensity distribution we would like to record causes similar optical property distribution in the medium. It can be either *transparency* of the recording medium (*amplitude hologram* is made) or its *optical thickness/index of refraction* (a *phase hologram* is made). Which one is relevant depends on the used light, its intensity and the kind of recording medium. The commonly known one is the photographic material. The photographic film gets darker where the original image was lighter. On that case, mostly the transparency of the medium is changed. Bleaching may transform it into a phase hologram.

3.2 Reconstruction of a hologram – what is hidden there?

While recording a hologram, no optical system to create the image of the object was used. This way, after processing the recording medium, nothing can be seen by naked eye. Only a microscope would show us a very tiny interference structure (maxims and minima).
A question arose — how to see what was recorded on the hologram? To understand, it might be better firstly to describe what to do to see what is hidden in the hologram, in another words - to reconstruct its content. Then we shall try to understand why it is this way.
To record a hologram of an object the interference of two waves (object wave and reference one) is recorded. In other words, an interference structure is recorded.

To reconstruct the hologram, certainly it is necessary to illuminate the hologram. That means to illuminate a structure. Now we meet the second important physical phenomenon as the principle of the holography – *diffraction* (*Fraunhofer's one*) of light reconstructing a hologram.

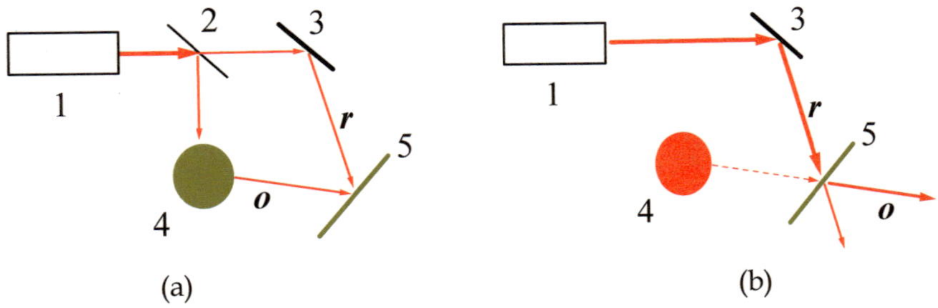

Fig. 9. Recording (a) and reconstruction (b) of a hologram (1-laser, 2-beam splitter, 3-mirror, 4-object, 5-hologram)

Fig. 9 demonstrates recording and reconstruction of a hologram schematically. A hologram had been created by interference of two waves ***r*** and ***o***. The interference structure was recorded. When illuminating the structure by one of two waves, which had created it, the second wave appears, too. Naturally, we would like to see which an object had been recorded and illuminate the hologram with the reference wave ***r***. Besides a new wave ***o*** appears, which is the same wave as it came from the object, and observer can see the object. Let us try to explain that "miracle" in a simply way.

Huygens' principle might be the simplest way. When recording a hologram, two waves overlap, interfere and create a resulting wave with special intensity distribution in all the space of overlapping. In one of the planes, the interference pattern is recorded.

When one of the waves creating the hologram (usually the reference one) illuminates the hologram the same intensity distribution as during the hologram recording appears just behind the hologram. We get the same point light sources distribution as during the interference of former object and reference waves. That means, following the Huygens' principle, – the same waves have to spread from them as before, i.e. the reference wave and the object wave. It is said – the object wave was reconstructed, object can be observed, again.

Let us show it a bit more exactly, using the complex notation (5) to express both the waves and the process of recording.

Interference of a reference wave $\boldsymbol{r} = R.\exp(i\phi_r)$ and the object wave $\boldsymbol{o} = O.\exp(i\phi_o)$ is recorded at a plane recording medium. The resulting amplitude ***a*** incident at the recording medium can be expressed by the sum $\boldsymbol{a} = \boldsymbol{r} + \boldsymbol{o}$. Only the intensity $I \sim \boldsymbol{a}.\boldsymbol{a}^*$ can be recorded

$$I \sim (\boldsymbol{r}+\boldsymbol{o}).(\boldsymbol{r}+\boldsymbol{o})^* = R^2 + O^2 + \boldsymbol{ro}^* + \boldsymbol{or}^* \tag{25}$$

Product of two complex conjugate numbers (like $\boldsymbol{r}.\boldsymbol{r}^*$) gives a real number equal to the square of absolute value of the relevant amplitude (R^2).

For simplicity, let us consider a photographic recording medium. When taking a proper exposure time and processing the medium properly, its *amplitude transparency* ***t*** is determined by the intensity (25) (Kreis, 2005). Amplitude transparency is defined as the relation of the amplitude $\boldsymbol{a}_t$ passing through the transparent (our hologram) and the amplitude ***a***, which illuminates the transparent, i.e. $\boldsymbol{t} = \boldsymbol{a}_t/\boldsymbol{a}$. Since dealing with amplitudes of waves, it cannot be measured.

We shall express what happens during reconstruction. The hologram is illuminated by the *reconstructing wave* f. When supposing a thin hologram, the wave f_t just behind the hologram can be expressed in the form

$$f_t = f.t \sim f.I \sim f.(R^2 + O^2) + f.r.o^* + f.o.r^* \tag{26}$$

Relation (26) shows that when illuminating the hologram with the reconstructing wave $f = r$, light field just behind the hologram consists of three parts. To understand the principle, it is not important to deal with all the parts in detail, now. Let us notice the last part, only, which can be expressed in the form

$$f.o.r^* = r.o.r^* = R^2.o \sim o \tag{27}$$

It has been proved exactly in a simply way that the object wave is included into the light field just behind the hologram, when illuminating it with the reference wave. Because of that we can see the object despite being not able to finger it. When using lens terminology, a *virtual image* is reconstructed (Fig. 10b).

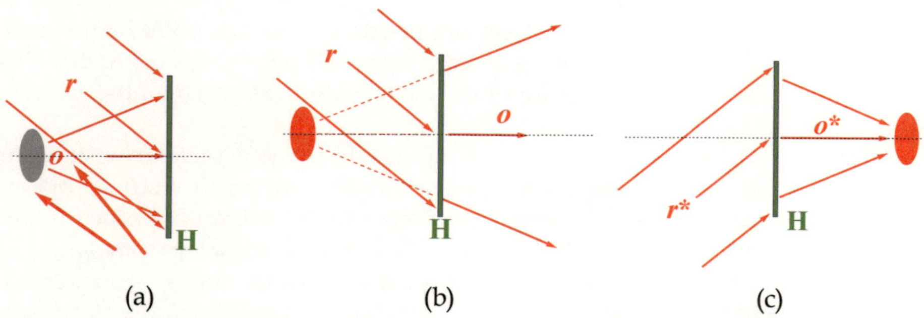

Fig. 10. Hologram recording (a) and reconstruction of virtual (b) and real (c) image

However, when illuminating the hologram with the complex conjugate reference wave $f = r^*$, the second part of (26) gives reconstruction of complex conjugate object wave o^*

$$f.o^*.r = r^*.o^*.r = R^2.o^* \sim o^* \tag{28}$$

Such a reconstructed image (Fig. 10c) can be touched and seen on a screen. A *real image* is reconstructed.

Both, lens and hologram create an image, so hologram is often compared to a lens. So-called *hologram formula*

$$\frac{1}{R_2^{(r,v)}} + \frac{1}{R_1} = \frac{1}{R_1}\left(1 + \frac{\lambda_2}{\lambda_1}\right) - \frac{\lambda_2/\lambda_1}{L_1} \pm \frac{1}{L_2} = \frac{1}{f_H^{(r,v)}} \tag{29}$$

similar to the *lens formula* can be derived, too. A hologram also can be characterised by its *focal length* f_H. It depends on the wavelength of light when either recording (λ_1), or reconstructing (λ_2), object distance from the hologram (R_1) and distance of the source of both, reference (L_1) and reconstructing (L_2) beam from the hologram (Kreis, 2005). Paraxial approximation is supposed, all the distances are measured perpendicularly to the plane of hologram and indices (signs) r (-), v (+) in (29) are related to real and virtual reconstruction. Concluding, let us mention that when speaking about holography using the language of diffraction, the reconstructed object wave is the diffracted maximum of the first order, created when reference beam is diffracted by the hologram.

3.3 Properties and types of holograms

There are many interesting properties of a hologram. Let us mention some of them. The first three ones, may be considered to be the most important when comparing a hologram to a photograph:

a. Holography is the only visual recording and playback process that can record our three-dimensional world on a two-dimensional recording medium and "playback" the original object or scene to the unaided eyes as a three dimensional image.

This property follows from what has been said above – the same wave, as propagating from the 3D scene/object while recording the hologram, is reconstructed. We can see the scene/object like through a window in the hologram size.

b. Excluding few special cases, the original scene or object can be reconstructed from a piece of the hologram, too.

Every point of the illuminated scene/object can be considered as a point light source from which a spherical wave propagates and can cover all recording medium. This way information about any illuminated point of the scene/object can reach all the surface of the recording medium (Fig. 11.).

The more is light scattered by the object the better is information spread over the hologram. However, only a part of hologram cannot be used to reconstruct all recorded object when the object does not scatter light, when light is reflected from a glittering surface. Only a part of hologram cannot be used also in the case of a kind of rainbow holography when a lens is projecting the object at the hologram. However, such a hologram can be reconstructed using white light and it will be told about later.

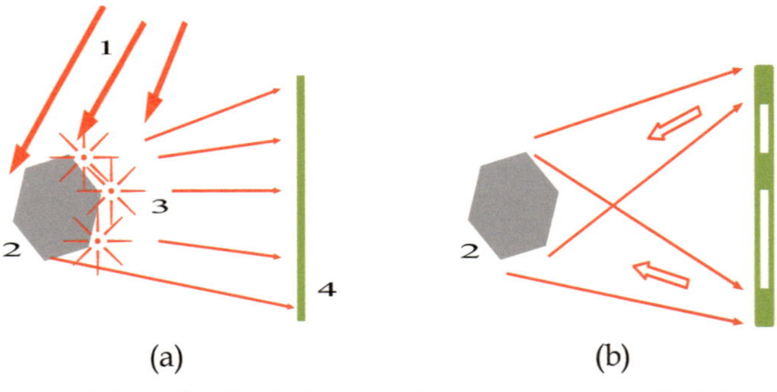

Fig. 11. a) Object wave 3 (set of spherical waves from point sources) of the object 2 surface illuminated by the wave 1; b) Every part of the hologram 4 shows everything

c. Many records can be made on the same recording medium and they can be reconstructed without interfering each other.

The holographic record is a record of a structure. It can be reconstructed only when proper orientation of the reconstructing (usually reference) beam towards the hologram is chosen. When making more records on the same recording medium, the orientation of the medium is gradually changed. However, the number of records is limited by the state when all the structures overlap so that they cannot be distinguished.

There are more other terms related to holograms to characterize them, like

d. *Plane* and *volume* holograms, which differ from each other by the relation between the thickness h of the recording medium and recording density following from relation (24). A hologram is said to be a volume (thick) one, when approximately $h \geq 1.6(\Delta x)^2/\lambda$. Thickness of recording medium decides for example about its storage capacity and possibility to be reconstructed using white light.

e. *Laser* or *white light* for reconstruction — any hologram can be reconstructed when using laser. However, there are also special holograms, which also can be reconstructed using white light. Some of them can be met daily e.g. at banknotes, and cards.

To understand the item, let us notice the physical meaning of reconstruction - diffraction of reconstructing wave by the "*grating*" of a hologram. For simplicity, let us consider a hologram like a simple grating with grating constant d and remember the condition (20) for grating m-order interference maximum

$$d.\sin\alpha_m = m\lambda,\ m = 0, 1, 2, 3, ...$$

where λ is wavelength and grating constant d is given by the difference between two adjacent interference maxims when the hologram had been recorded (24). When considering reconstruction, m =1. For simplicity, the index m is not used later. Since the angle α depends on the wavelength λ various angles $\alpha(\lambda)$ give various positions of the reconstructed object (Fig. 12.) Naturally, it results in "blurred" reconstructed object. However, looking at Fig. 12, it could occurred to us that if an object closer to the recording medium had been recorded, a less blurred reconstruction could have been obtained. There is a possibility to put the recorded object even "into" the recording medium and not to restrict the reference wave while recording the hologram - to *project* it there by a lens/objective. When reconstructing in white light the reconstruction is observed in various (rainbow) colours from various directions at the same place (Fig. 13.).

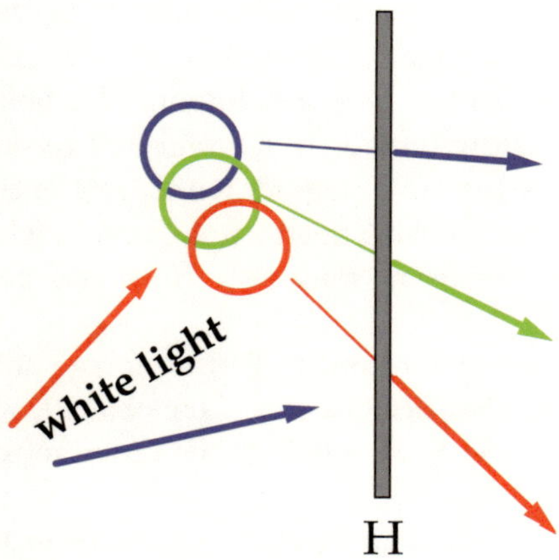

Fig. 12. White light reconstruction

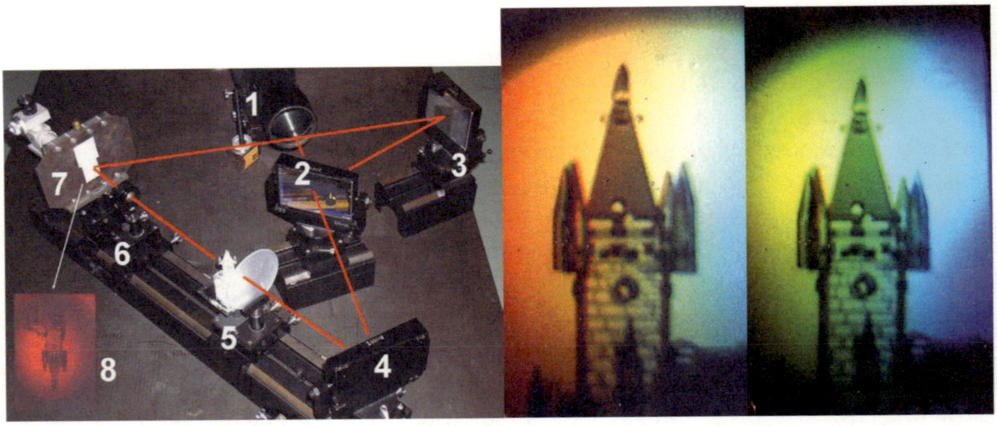

Fig. 13. Experimental set-up and white-light reconstruction of a rainbow hologram with a projecting element (1-from laser, 2-beam splitter, 3 and 4-mirrors, 5-ground glass and object, 6-projecting lens, 7-recording medium, 8-image of the object)

Naturally, projection of a 3D object has 3D properties, too. It is said that a human eye cannot distinguish blurring, caused by the "thickness" of the projection of about 1cm. Because of that, such a simple method is used mostly for approximately "2D" objects (Fig. 14.) - medals, coins, photos, and so on.

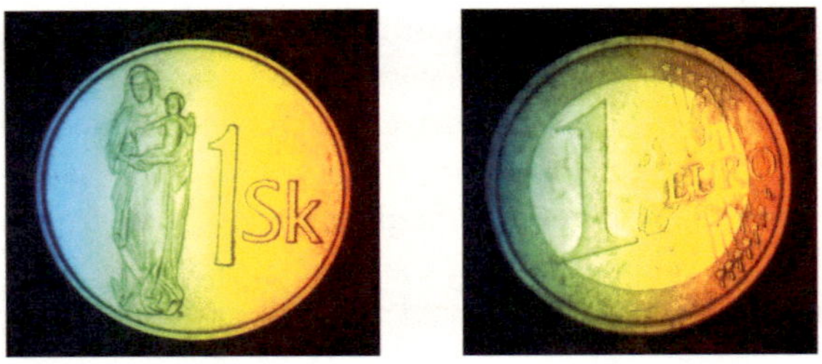

Fig. 14. Slovak crown and euro

When a real 3D object is recorded the well-known Benton's method (Benton, 1969), which allows us to get rid of vertical blurring, has to be used. The Benton (rainbow) hologram is a transfer transmission hologram. In fact, it is a hologram of a hologram. The first (master) hologram is masked with a narrow horizontal slit and real reconstruction of its free part serves as an object to record the second hologram with a reference wave diverted from the object one in vertical plane. When viewing such a hologram in white light the colour of the hologram changes and hence the term "rainbow". However, perspective information in vertical axis is lost.

f. *Fourier* hologram is a holographic record of Fourier image of a 2D object. In its second focal plane the lens 2 creates Fourier image 3 of the object 1, which is placed in the first focal plane of the lens (Fig. 15.). Such a hologram is used mostly in optical information processing.
g. *Denisjuk's* hologram is another hologram, which can be reconstructed in white light. On the contrary to thin holograms, observable in transmitted light, it belongs to *thick* holograms observable in reflected light.

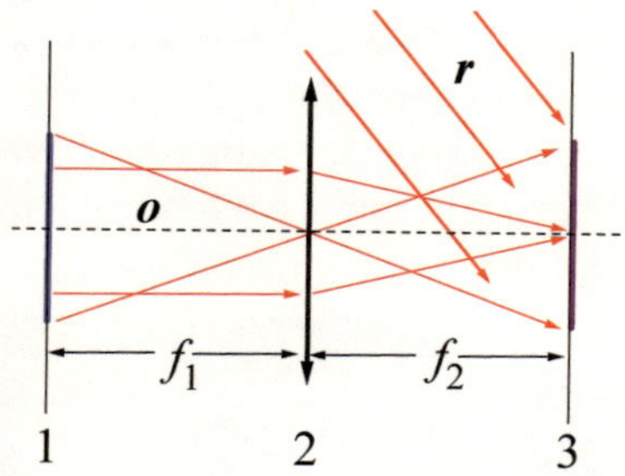

Fig. 15. Fourier hologram

The idea based on Lippman's colour photography (Denisyuk, 1962), had been elaborated by a Russian physicist Yu. Denisyuk (Fig. 17). After lasers became available Denisyuk developed *volume reflection holography*. Denisyuk's technique was different in concept and implementation from Gabor's one. His method could reconstruct three-dimensional holographic images by reflection from the hologram in white light. Denisyuk presented his technique as a generalized form of Lippmann photography, or as a color-dependent optical element. This technique, using reflection holography and the white-light reconstruction technique, seems to be the most promising one as regards the actual recording of colour holograms (Bjelkhagen, H. I. in Ludman et al., 2002).
Considering basic optics it is a kind of a volume grating. Ideally, in such a case only one first-order diffraction maximum is observed, if well-known Bragg's condition (Kreis, 2005)

$$2dn.\sin\alpha = \lambda \tag{30}$$

connecting the grating constant d, wavelength λ and angle α between incident/reflected beam and plain of grating is fulfilled. Index of refraction of the recording medium is n. In other words – the hologram interference structure can choose a proper reconstructing wavelength from incident white light according its direction and the reconstructed object is observed in monochromatic light.

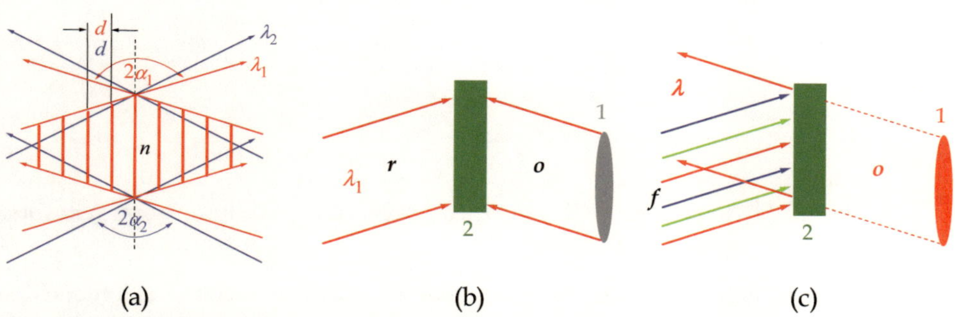

Fig. 16. Thick hologram and white light reconstruction (1-object, 2-hologram)

Fig. 16 shows it schematically. Taking into account relation (24), two coherent plane waves (λ_1) with the angle $2\alpha_1$ between each other, create the interference structure with the grating constant d in medium with index of refraction n

$$d = \lambda_1 / 2n\sin\alpha_1 \tag{31}$$

which is identical with relation (30). When another waves (λ_2), another angle $2\alpha_2$ has to be used to create an interference structure with the same grating constant d (Fig. 16a).
Let us create hologram 2 of the object 1, using light with λ_1 (Fig. 16b). When such a hologram is illuminated by white light f in the reference wave λ_1 direction only wave λ_1 is able to reconstruct the object (Fig. 16c).

Fig. 17. Yu. Denisyuk reconstructed from a hologram and alive (VRC, 2006)

However, as mentioned above, such a reconstruction is a monochromatic one, only. What about colour holograms? To produce colour hologram, one needs three lasers generating on three basic wavelengths (red, green, blue) to record such a hologram. Each of the waves creates own interference structure (Fig. 18). After illuminating such a hologram with white light, each structure helps to reconstruct the object wave in related wavelength. The reconstruction is seen in three colours and thanks to sophisticated activity of our brain as colourful.

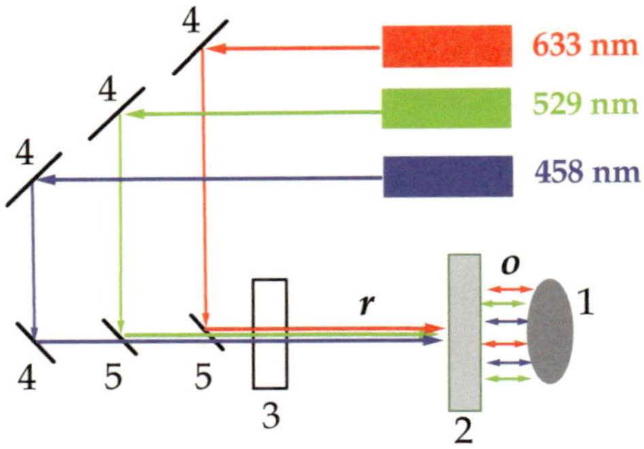

Fig. 18. A colourful hologram recording (1-object, 2-recording medium, 3-optics to spread light, 4-mirror, 5-semitransparent mirror

h. Well, not to forget, something interesting, else – to perform reconstruction one does not need to use the same wavelength (λ_2) as while recording (λ_1). When $\lambda_2 > \lambda_1$, the reconstructed 3D image will be magnified. Just that was the marvellous idea of Dennis Gabor: *... to record the hologram in the region of X-rays (~10^{-10}m) and to reconstruct it in the visible (~10^{-7}m) region, great magnification (~λ_2/λ_1) can be obtained without any limitation by objective properties....* (Gabor, 1971).

Both, transverse $\beta_\perp$ and longitudinal β_L magnification (Kreis, 2005, pp. 44-47) depends on recording and reconstructing set-ups (R_1, L_1, λ_1, L_2, λ_2)

$$\beta_{\perp}^{(r,v)} = \left[1 - \frac{R_1}{L_1} \pm \frac{\lambda_1 R_1}{\lambda_2 L_2}\right]^{-1} \qquad \beta_{L}^{(r,v)} \approx \frac{\lambda_2}{\lambda_1}\left[\beta_{\perp}^{(r,v)}\right]^2 \tag{32}$$

where sign −/+ is related to real (^{r})/virtual (^{v}) reconstruction.

3.4 Some applications

Great development and application of holography began after invention of laser producing coherent light, which is necessary to produce a hologram. Let us mention briefly some examples, at least. The explanation is simplified and only the basic ideas are pointed.

3.4.1 Holographic storage

Storage requirements are exploding today. Faster access, higher data rates and redundant storage of data within the volume of a thicker medium require a new approach. Optics and the basic principles of holography are a particularly attractive way to meet these requirements and may well represent the storage solution of the future. Holographic storage represents an opportunity to significantly increase data densities beyond those offered in conventional removable storage technologies, and to increase data transfer rates well beyond those that might be envisioned from today's storage products.
Data are encoded onto the object beam by spatial light modulator (SLM), which translates the electronic data of 0s and 1s into an optical "checkerboard" pattern (Fig. 19) of light and dark pixels. Unlike other technologies holographic one writes data through the full depth of the recording medium. By varying the reference beam angle or media position hundreds of unique holograms are recorded in the same volume of material. The effective area storage density (bits/unit area) can be significantly increased by using a thick recording layer to record multiple, independent pages of data. The holographic structure for one page is intermixed with the holographic structures of each of the other pages. This process has been known as *multiplexing*.

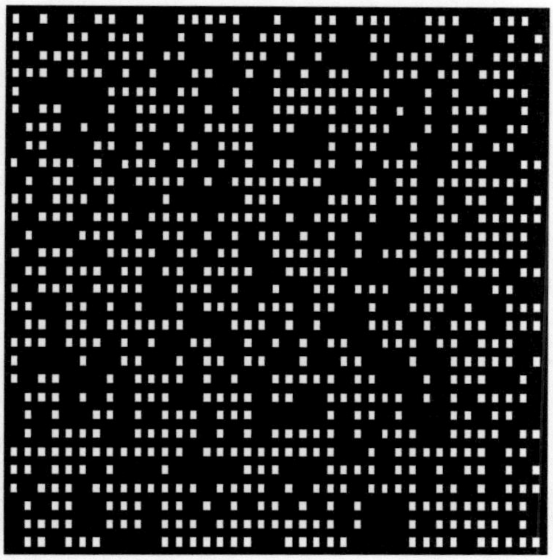

Fig. 19. 2D digital data

Retrieval of an individual page with minimum cross talk from the other pages is a consequence of the volume nature of the recording and its behaviour strongly depends on mismatches in angle or wavelength between recording and reconstruction. Moreover, information distribution throughout the recording volume allows reducing sensitivity to material defects.

It is most important to realise that over a million bits of data are written and later read with a single flash of light holographically. Namely, data pages in whole and not serial stream of bits, only, are either written or read simultaneously.

Since holography makes use of the full thickness of the recording material, providing data densities proportional to media thickness, capacities of more than 1,000 GB on a CD disk format, can be achieved.

Data stored holographically are transferred as pages of optical information. This parallel read out of data provides holography with its relatively fast transfer rates. Consequently, holography provides a substantially faster data transfer rate from a single head, surpassing 100 MB/sec.

Companies developing holographic data storage systems have made enormous technical advances towards their goal of inventing an optical data storage system (InPhase, 2007; STX Aprilis, 2006). A prototype that holds 200 gigabits per square inch of storage capacity was demonstrated in 2007.

3.4.2 Optical processor and holographic optical elements

Generally, optical processor is a device used to process optical information, i.e. to record, to retrieve and to recognize it. Fig. 20 demonstrates the principle of any optical processor – projection utilizing two lenses 2 and 4 with common focal plane 3. The lens 4 projects the object from the first focal plane of the lens 2 into its second focal plane. Why such a complicated projection? The common focal plane 3 becomes accessible to filtrate optical information processed. Namely, it is the plane where Fourier transform (Smith et al., 2007) of the object is created.

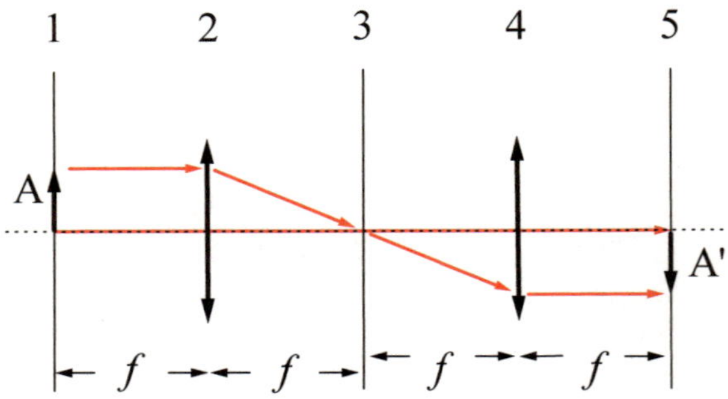

Fig. 20. Optical processor (2, 4 - lenses, 1 -object plane, 3-filtration plane, 5-image)

In principle, the experimental set-up of a real optical processor follows the set-up used to record and retrieve a Fourier hologram. It is only equipped with helping elements, enabling multiple recording and reading. However, the object has to be in form of 2D transparent covered by pattern of 1s and 0s and plane wave is used as a reference one.

Especially, information recognition process, when we are looking for a certain kind of information (e.g. A) in noise, takes a great attraction. A proper filter, Fourier hologram of A - *FH*(A) has to be prepared (Fig. 21a). Now, the object wave $\boldsymbol{\Sigma o}$ in which $\boldsymbol{o}_{\mathbf{A}}$ is supposed to be included enters the optical processor.

When there is no filter in the common focus plane 3, image of the object related to the wave $\boldsymbol{\Sigma o}$ is created in the output focal plane 5 of optical processor. If information A is included in $\boldsymbol{\Sigma o}$, a plain wave occurs in light of Fourier transform of $\boldsymbol{\Sigma o}$ transmitting the filter *FH*(A). After passing through the second lens it creates an intensive light point R in its second focal plane (Fig. 21b). The appearance and position of this bright point tells us about presence and position of information we are looking for in noise.

Naturally, Fig. 21 is simplified. In reality, one does not know the position of A and *FH*(A) is made with A on the axis. It only results in a shift of the image in whole with the light point *R* in the output focal plane 5.

Where to use such a method? What about fingerprints recognition in a database of fingerprints used instead of noise with information included?

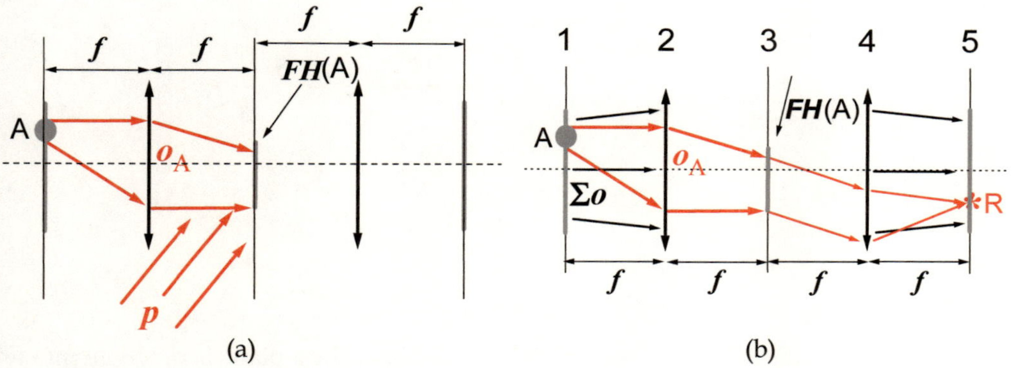

Fig. 21. Information recognition - making a holographic filter to recognize information A (a) and recognition A (b)

Holography also enables us to create various optical elements, like lenses, gratings, prisms, beam splitters, and so on, in a simply way. Despite working by diffraction and not by reflection or refraction, they can be used for any purpose that conventional optical elements can be used for.

Especially holographic gratings can demonstrate the advantage of such an approach. To make classically the dense line structure of a grating, a special ruling engine had to be used. Instead mechanically ruled grating grooves a hologram of a plane wave with plane reference wave can be recorded. The interference pattern has the form of regular parallel strips - intensity maxims and the grating interval can be changed very simply, by choosing a proper angle between the interfering waves (24). Besides spectroscopy, a grating can be used as either a beam splitter or a beam-directing element.

Similarly, a hologram of a converging or a diverging wave can be used as a lens.

3.4.3 Holographic interferometry

Interferometry is a method based on interference of light. It enables to determine the object variations in the scale of the wavelength λ of used light. Such a method had been known and used since man got familiar with the phenomenon of interference of light.

To observe interference coherent light is required. Because of that invention of lasers, sources of coherent light became a great advancement. Classical interferometry had been limited by the high optical quality requirements for both all the optical elements used in the set-up and studied surfaces (roughness $< \lambda/20$).

Holography brought something impossible to be performed before. Holography enabled to study rough surfaces, which scatter incident light. Moreover, it added a kind of bonus – to examine objects, even not existing already. A simply explanation can be found using Fig. 22, where basic experimental set-ups used for classic (a) and holographic (b) interferometry are shown.

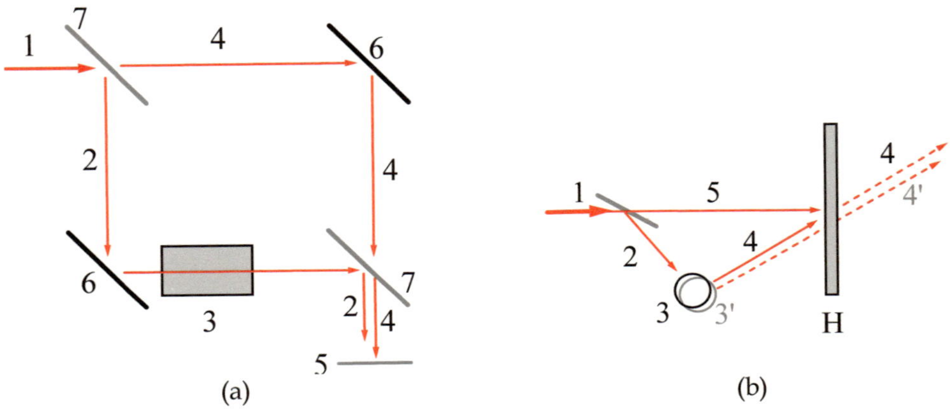

Fig. 22. Classical (a) and holographic (b) interferometry

In classical interferometry (a) the original light wave 1 (usually a plane one) is split into two parts. One of them (2) passes the examined object 3 (transparent one in our case). The second wave 4 spreads without any change. Both waves are directed by splitter 7 and mirrors 6 mostly to the same direction to cover each other. They are coherent and able to interfere. The interference pattern can be observed/detected/recorded in any plane like 5. The local resulting intensity of light depends on the phase difference between two waves met at that point. Any change of the object causes a relating change of the interference pattern shape and intensity, which allows us to estimate the changes of the object.

Naturally, all of that is true *only* when the changes of the wave 2, comparing to the wave 4, are caused *only* by the changing object. Just that is the reason of high optical quality of all the elements and object surfaces required. The wave changed by an object is compared to a kind of "an ideal" wave.

In holographic interferometry (b), at least one of the interfering waves has to be reconstructed from a hologram. An original wave 1 is split into two parts, again. One of them is scattered by the object 3 (an example of non transparent object is used now) and creates the object wave to record a hologram. The second part serves as a reference wave 5 to record the hologram of the object in a basic state. There are more ways how to proceed now. For example, the hologram is recorded, processed and returned back, into the same set-up and place where recorded — so-called *real-time interferometry*. Now, reference beam 5 reconstructs the object wave in the basic state. The changing object provide us with the next object wave 4', which interfere with the reconstructed wave 4. Interferogram varies in real time, simultaneously with the varying object.

As you may have noticed, now, the changing object wave does not interfere with a kind of "an ideal" wave. The interference pattern results now from the interference of two object waves, which differ due to various states of the object. They represent two different states of the object. The interference observed is not handicapped by any possible low optical quality of the set-up elements.

The reason is that the low quality is *the same* in both states of the object. Two interfering waves 4 and 4' differ *only* because of changing object.

Another possibility would be for example — to record both the states of the object without any other change in the set-up, to process the hologram and reconstruct two object waves simultaneously. It is said to be *double exposure holographic interferometry*. When taking into account the pros and cons, it is very useful especially in the case of fast running processes.

Because of the interference of two object waves holographic methods in interferometry enable to study objects scattering the incident light. Such objects are impossible to be studied in the frame of so-called classical interferometry.

And why could we examine objects and study interferograms even without having a real object? Of course, we have to have it when recording holograms. We can have holograms of various states of the object and reconstruct them two various at once. Two reconstructed waves of two various states of the object interfere and create the interference pattern one can study.

Concluding, I would like to illustrate an example of holographic interferometry used to find optical thickness radial profile of an optical fibre (Fig. 23).

Fig. 23. Interferogram - holographic interferometry of an optical fibre

A plane wave crossing the fibre served as the object beam. The fibre was projected at the hologram with a proper magnification, double exposed hologram was recorded and the phase shifting interferometry was applied, i.e. the angle of incident of the reference beam was a little bit shifted before exposition of the second state of the object. As two "object states were used a cell filled with glycerine and the fibre embedded into it and the same cell filled with glycerine without the fibre. This way the producer's data, relating to the preform were approved. The paper "Senderakova et all. (2004), Interferometric analysis of optical fibre profile" was published in Slovak, only, so it is not included into references.

4. Conclusion

I would like to notice that information above has not been any complete and exhaustive overview of the field of holography. Internet can show itself as a rich source of simple explanations as for holography and its applications. Material collected and submitted here, should only help to understand the basic physical principles used in holography. It may open a gate to understanding various interesting and many times astonishing applications of holography one can meet today.

When understanding the principle, a hologram can be generated using a computer, too. *Computer-generated holograms* find applications especially in the role of holographic optical elements. Method of electron beam lithography enables to write structures on the order of 0.1 μm. Such holograms are reconstructed optically.

Digital holography is another contemporary term. In this case, hologram is recorded and stored electronically. Unquestionably, it is much simpler way comparing to using a photo-material. However, there is a restriction - size of pixels, which has to be taken into account, when regarding spatial frequency of holographic interference pattern (Kreis, 2005). When using wavelength λ, the spatial frequency $1/d$ (24) is defined by the angle between object and reference waves. Moreover, when considering so-called *sampling theorem* (Kreis, 2005), i.e. the period d must be sampled with more than two pixels, the allowed angle between object and reference waves must not exceed approximately (2-3)°. Reconstruction, i.e. the diffraction pattern is calculated numerically (Schnars & Jueptner, 2005) in a computer. Digital holography finds its applications in interferometry. However, non-interferometric applications can be met, too - e.g. particle analysis, microscopy, and data encryption (Kreis, 2005).

Holography is not restricted to optical coherent waves, only. It also might be interesting to notice holography based on either acoustic or ultrasound waves - *acoustic holography* and *ultrasound holography*.

Electron holography is the application of holography techniques to electron waves. Let us realize that it was just electron holography, which was invented by Dennis Gabor, having been worked on improving the transmission electron microscope. The principle of such a holography can also be applied to interference lithography.

SPIE's 2011 International Symposium on Optics & Optoelectronics (Prague 2011) focused besides other technologies, also on holography. Sessions of one of the technical conferences during the symposium — *Holography: Advances in Holography and Modern Trends* (8074) shows contemporary topics in holography: digital and computer generated holography, security holography and holographic diffractive optics, recording materials and information storage, and holographic methods and other applications. For example, let us mention

- digital holographic microscopy, a non-invasive technique for imaging transparent samples
- computer generated holographic optical elements to compensate aberrations

- special photographic emulsion, organic glasses, photopolymers, photorefractive materials as recording materials
- polarization holography for optical trapping and manipulation, hologram multiplexing methods, holographic lithography for making large photonic crystals easily, application of the laser analyzer for identification in real time of security holograms for various documents, security holograms, optical holography and its applications in solar energy concentrations, and so on.

When asking about the future of holography, allow me to quote Emmett Leith, one of holography fathers (Ludman et al., 2002): *"Holography in the 21st century will continue to flourish. Its growth will result in large part from the advancement of the technologies on which holography depends: the computer, the electronic camera, the advances in real-time recording and display media, the advances in development of mask making and others. At the same time, as holography grows and as new forms develop, the boundary between holography and non-holography will become more indistinct."*

5. Acknowledgment

I would like to express this way my sincere thanks to Faculty of Mathematics and Physics, Comenius University at Bratislava, which provided me with a possibility to get familiar with beauty of new, contemporary optics. I, also, would like to express my thanks to all of my close colleagues and students, for transforming the atmosphere of my working place into a home atmosphere, indeed.

6. References

Benton, S.A. (1969). Hologram Reconstruction with Extended Incoherent Sources, *J. Opt. Soc. Am.*, Ser. A, vol. 59, pp. 1545-1546

Davidson, M. W. & The Florida State University. (1995). Timeline in Optics, In: *Molecular Expressions – Optics*, (March 2011), Available from: <micro.magnet.fsu.edu/optics/timeline/index.html>

Denisyuk, Yu. N. (1962). On the reflection of optical properties of an object in the wave field of light scattered by it, *Doklady Akademii Nauk SSSR*, vol.15, No 5, p.1275

Gabor, D. (1971). Holography 1948-1971, In: *The Nobel Prize in Physics 1971*, March 2011, Available from: <nobelprize.org/nobel_prizes/physics/laureates/1971/gabor-lecture.pdf>

InPhase. (2007). Overview, In: *InPhase Technologies: Products*, March 2011, Available from: <www.inphase-tech.com/products/default15b9.html?tnn=3>

Kreis, T. (2005). *Handbook of Holographic Interferometry (Optical and Digital Methods)*, WILEY-VCH Verlag GmbH & Co. KGaA, ISBN 3-527-40546-1, Weinheim

Levis, G. N. (1926). The Conservation of Photons, *Nature*, vol. 118, Part 2, pp. 874-875), Available from: <www.nobeliefs.com/photon.htm>

Ludman, J., Caulfield, H. J., & Riccobono, J. (Eds.). (2002). *Holography for the New Millenium*, Springer-Verlag New York, Inc., ISBN 0-387-95334-5, New York

Schnars, U., & Jueptner, W. (2005). *Digital Hologram Recording, Numerical Reconstruction, and Related Techniques*, Springer-Verlag Berlin Heidelberg, ISBN: 978-3-540-21934-7

Smith, F. G., King, T. A. & Wilkins, D. (2007). *Optics and Photonics* (2nd edition), John Wiley & Sons, Ltd, ISBN: 978047001838 (HB), Great Britain

Stevenson, D., C. (1994). Aristophanes – The Clouds, In: *The Internet Classics Archive,* (March 2011), Available from: <classics.mit.edu/Aristophanes/clouds.html>

STX Aprilis. (2006). Holographic Data Storage, In: *Media Products,* (March 2011), Available from: <http://stxaprilis.com/datastorage.htm>

VanRenesse Consulting. (2006). Yuri Denisyuk - Pioneer of modern holography, In: *Van Renesse Consulting – Photography,* March 2011, Available from: <www.vanrenesse-consulting.com/index.php?page=photography.htm>

2

Digital Holography: Computer-Generated Holograms and Diffractive Optics in Scalar Diffraction Domain

Giuseppe A. Cirino[1], Patrick Verdonck[2], Ronaldo D. Mansano[3],
José C. Pizolato Jr.[1], Daniel B. Mazulquim[4] and Luiz G. Neto[4]
[1]*CCET - Universidade Federal de São Carlos,*
[2]*IMEC,*
[3]*EPUSP – Universidade de São Paulo,*
[4]*EESC - Universidade de São Paulo,*
[1,3,4]*Brasil*
[2]*Belgium*

1. Introduction

The invention of holography by D. Gabor, followed by the works of E. Leith and Upatnieks, made possible to perform nearly arbitrarily wavefront transformations with the aid of optical microstructures fabricated by interference (Gabor, 1948; Leith & Upatnieks, 1962; Denisyuk, 1962; Benton, 1969). The realization by A. Lohmann and co-workers in the mid-1960's that optical holograms can be simulated by digitally generated binary transparencies was another significant step forward in the path towards widespread application of diffraction in optics (Lohmann, 1956; Brown & Lohmann 1966; Lohmann & Paris, 1967). This new approach was called digital holography, or equivalently computer-generated holography. The computer-generated hologram (CGH) is distinguished primarily from its optical cousin by the fact that the computer is able to design a hologram of a non-existent, synthetic or virtual object, and the operation of the diffractive optical element (DOE) can be optimized mathematically rather than experimentally (Turunen & Wyrowski, 1997). This is clear from the words of Professor Lohmann himself, commenting on the shift from classic holography to digital holography (Lohmann, 2008):

> "we considered images as information, and we applied notions about carriers from communications and information theory [...]. In other words, our approach represented a paradigm shift from physical optics to optical information processing. "

Since the operation of CGH is based on the diffraction of light, this field is also called diffractive optics. Its essence is the control of optical fields by microstructured media (Turunen & Wyrowski, 1997).

The DOE is an optical device whose superficial microrelief has a height comparable to the light wavelength used. The DOE may be implemented in the form of a transparency or a reflecting mirror.

Throughout this text the terms computer-generated holograms (CGH), diffractive optical element (DOE), or simply hologram are employed with no distinction among each other.

The goal of digital holography is to form a light distribution in the observation (or reconstruction) plane. There are many successful applications in advanced scientific and technological fields, such as optical lithography and fabrication, and photonic manipulation of particles (optical tweezers). Lenses, zone plates, diffraction gratings, array illuminators, kinoforms, and the phase spatial filters are other examples of DOEs (Soifer et al., 1997).

The technology of DOE manufacturing involves microfabrication processes which have been maturated from microelectronics and MEMS/MOEMS technologies. Once the design is completed, an accurate process is used to materialize the DOE, such as laser, electron or ion beam writing, half-tone mask technique, diamond turning, and so on. A broad list of micromachining approaches is described in references (Herzig 1997; Turunen & Wyrowski, 1997). For a low-cost, mass production, of surface-relief microstructures, replication techniques must be used. It includes technologies such as hot embossing, casting and injection molding (Herzig 1997; Turunen & Wyrowski, 1997).

This chapter covers some aspects of design and fabrication of CGHs, operating in the scalar diffraction domain, implemented in a multi-level surface-microrelief. The specific contributions of the presented research, is the implementation of a full complex-amplitude modulation CGH, sections 3.2-3.5. Its design is based in a sub-cell approach, which increases the degree of freedom, enabling high quality diffraction patterns. Another contribution - in the fabrication of such a diffractive element - is the employment of an amorphous hydrogenated carbon (a:C-H) thin film, also known as Diamond-Like Carbon (DLC). A patented reactive magnetron sputtering technique used to produce the DLC thin film, proved to be very suitable for diffractive optics applications. Examples of amplitude-only, phase-only and complex amplitude modulation CGHs, operating in both Fraunhofer and Fresnel regimes, are presented (Neto et al, 2001, 2003, 2004, 2008).

2. Design of computer-generated holograms

The design of CGHs can be divided into three basic stages (Mait, 1995): (1) understand the physics of the design problem (analysis), (2) translate the physical understanding into mathematics and define an appropriate optimization problem (synthesis), and (3) execute the design and fabricate the element (implementation).

In the second stage one has to decide if scalar diffraction is enough to solve the problem accuratelly or it demands rigorous analytic solutions based on vectorial electromagnetic theory. Generally, rigorous analytical solution is difficult to obtain and tend to be computationally time-consuming (Turunen & Wyrowski, 1997). For most applications the employment of scalar diffraction theory is enough to represent practical engineering solutions (Mait, 1995). The analysis of DOEs by the employment of rigorous vector diffraction theory is beyond the scope of this chapter.

A DOE is designed using computer calculations based on the scalar diffraction of light, on the characteristics of the material (media) where DOE is to be fabricated, on the desired light distribution to be generated and determined by the fact of generating a phase-only or an amplitude-only distribution (object). To assure scalar domain, a linear dimension that reflects the smallest feature of the microrelief, D_{min}, must be at least ten times larger than the operating wavelength, λ. In the examples shown in section 3, the minimum feature size is *10X10* μm, and the diffractive elements operate at *632.8* nm wavelength He-Ne laser. Therefore the inequality above holds: $(10/0.633) = 15.8$.

Light, considered in terms of the scalar diffraction theory, being monochromatic and coherent, is described by a complex function of two spatial variables. The propagation of light in free space can be described using the Kirchhoff diffraction approximation, and the mathematical aspect of light propagation through an optical element reduces to the multiplication of the complex function of incident light by the complex transmittance of the optical element (Soifer et al., 1997).

Calculate a CGH means the calculation of its complex transmittance. This transmittance is expressed as $g(x',y') = A(x',y').exp[j\varphi(x',y')]$, where $A(x',y')$ is its amplitude distribution, $\varphi(x',y')$ is its phase distribution, and (x',y') are coordinates in the frequency space.

The phase modulation can be obtained by a microstructured surface of an optical medium. Amplitude modulation can be obtained by selectivelly blocking/attenuating the light intensity along the device's surface.

Depending on the media where the DOE is to be fabricated, one can refer to

i. phase-only DOE: When such a media is able to modulate only the phase of an incoming wavefront, the transmittance amplitude being equals unity;
ii. amplitude-only DOE: When such a media is able to modulate only the amplitude of an incoming wavefront, the transmittance phase being equals to a constant, or
iii. complex-amplitude DOE: When such a media is able to modulate both the phase and amplitude of an incoming wavefront.

To generate the reconstruction, it is possible to choose between designing a Fourier DOE, which employs Fraunhofer diffraction for the calculation of the element and a Fourier lens to implement the optical reconstruction, as shown in figure 1a, or a Fresnel DOE, by considering only the free space propagation of the light to generate the optical reconstruction, as shown in figure 1b.

If the element is illuminated by a uniform monochromatic coherent plane wave with a constant spatial amplitude distribution equals to 1, the light distribution just after the element has the form $1\times g(x',y') = g(x',y')$. The light propagation of the distribution $g(x',y')$ results in the optical distribution $g(x,y)$, located in the reconstruction plane, figure 1.

In the calculation of a DOE, the following design issues must be considered:

- Determination of the light wavefront modulation distribution $g(x',y')$ that should be generated in the DOE plane. The calculation of this distribution starts from the inverse light propagation of the desired light distribution $f(x,y)$ that should be generated in the reconstruction plane, back to the DOE plane. The main goal in the design is to determine the DOE distribution $g(x',y')$ that generates a reconstruction $g(x,y)$ as close as possible to the desired distribution $f(x,y)$, with $0 \leq |f(x,y)| \leq 1$. This light distribution $f(x,y)$ should be described in such a way as to permit computational manipulation, generally in a discrete form (usually discrete values of amplitude).
- The resulting inverse light propagation distribution $f(x',y')$ usually presents values of both phase and amplitude, which means that this distribution must be adapted to the possible DOE distribution $g(x',y')$, which can be a phase-only, an amplitude-only or a complex-amplitude distribution. The physical and practical limitations of the DOE medium will determine the choice of the numerical method used for its implementation.
- A distinction is made between complex amplitude objects and intensity objects (only amplitude determined). In this manner, for complex amplitude objects one has: $f(x,y) = |f(x,y)|\exp[\varphi(x,y)]$, and for intensity $i(x,y)$ objects: $|f(x,y)|^2 = i(x,y)$, where the argument value $\varphi(x,y)$ of $f(x,y)$ has total freedom and could assume any value, hence: $f(x,y) = \sqrt{i(x,y)}\exp\left[i\varphi(x,y)\right]$.

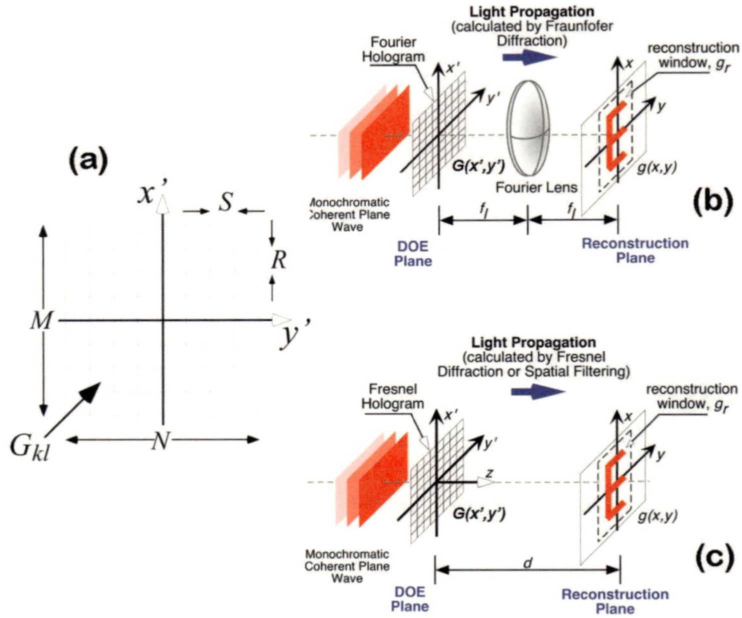

Fig. 1. (a) Hologram structure; (b) Optical reconstruction of Fourier-type CGH illuminated; (c) Reconstruction of a Fresnel-type CGH, obtained by the free space light propagation of $g(x',y')$ along distance d. Reprinted with permission of SPIE (Neto, L.G. 2004).

The hologram is considered a two-dimensional matrix structure with M by N (M and M being even numbers) with rectangular cells of dimension R by S. Each cell represents a complex transmittance value G_{kl}. Supposing the hologram is placed in the frequency plane:

$$G(x',y') = rect\left[\frac{x'}{MR},\frac{y'}{NS}\right]\left\{\sum_{k=-M/2}^{+(M/2)-1}\sum_{l=-N/2}^{+(N/2)-1} G_{kl} \times rect\left[\frac{x'-kR}{R},\frac{y'-lS}{S}\right]\right\} \tag{1}$$

Calculating the inverse Fourier transform of equation 1, one obtains the light distribution in the image formation plane:

$$\begin{aligned} g(x,y) &= M^2 RS \text{ sinc}\left[MRx, NSy\right] \otimes \\ &\left\{\sum_{k=-M/2}^{+(M/2)-1}\sum_{l=-N/2}^{+(N/2)-1} G_{kl} \exp\left[j2\pi\left(Rxk + Syl\right)\right] \times \text{RS sinc}\left[Rx, Sy\right]\right\} \end{aligned} \tag{2}$$

where $\otimes$ denotes the convolution operation. The signal $f(x,y)$ is also represented by discrete values: $f(x,y) \leftrightarrow f_{mn}(x,y)$.
Sampling the function $g(x,y)$ with intervals X in the x direction $[x=mX=m/(MR)]$ and Y in the y direction $[y=nY=n/(NS)]$, one obtain (Goodman, 1996)

$$g(x,y) = \text{sinc}\left[MRx, NSy\right] \otimes \left\{ g_{mn} \sqrt{\frac{\text{RS}}{\text{XY}}} \text{ sinc}\left[\frac{m}{M},\frac{n}{N}\right]\right\} \tag{3}$$

where g_{mn} is the discrete inverse Fourier transform of G_{kl}:

$$g_{mn} = \frac{1}{\sqrt{MN}} \left\{ \sum_{k=-M/2}^{+(M/2)-1} \sum_{l=-N/2}^{+(N/2)-1} G_{kl} \exp\left[j2\pi \left(\frac{mk}{M} + \frac{nl}{N} \right) \right] \right\} \tag{4}$$

The discrete Fourier transform g_{mn} is obtained numerically by using the Fast Fourier Transform Algorithm (FFT) realized by (Cooley & Tukey, 1965).

2.1 Iterative Fourier Transform Algorithm (IFTA)

The FFT algorithm gives a computational basis for virtually all of iterative methods (Soifer et al., 1997). In many cases, an error reducing algorithm called Iterative Fourier Transform Algorithm (IFTA), depicted in figure 2, is a natural choice (Wyrowski, 1989, 1990).

In the IFTA, one iteration is achieved by firstly calculating the inverse light propagation from the reconstruction plane to the hologram plane, where the hologram restrictions are applied (to force a phase-only distribution, as exemplified in figure 2), and then calculating the forward light propagation from the hologram plane to the reconstruction plane, where restrictions are applied (forcing the desired intensity reconstruction $i(x,y)$). For intensity objects, a random phase distribution is attributed to $\varphi(x,y)$ in the first iteration to help in the convergence of the algorithm.

Usually these numerical methods are computer time consuming and they introduce speckle noise in the reconstruction plane. A hundred or more iterations are needed, depending on the complexity of the design and on the desired light distribution $f(x,y)$ in the CGH reconstruction plane.

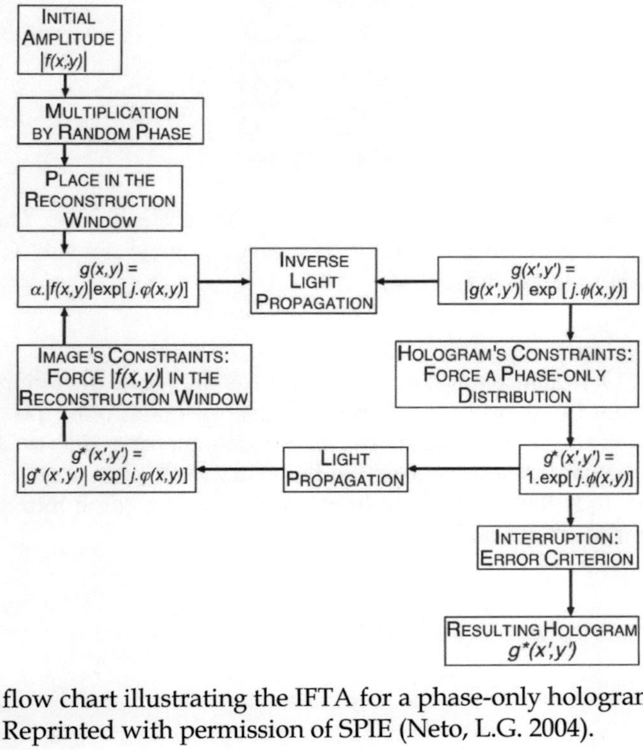

Fig. 2. Schematic flow chart illustrating the IFTA for a phase-only hologram and for intensity objects. Reprinted with permission of SPIE (Neto, L.G. 2004).

2.2 Phase, amplitude and scale freedoms

To help the convergence of the numerical method, some additional degrees of freedom can be introduced in the design. Assuming, for example, a phase-only CGH is to be designed (figure 2), they can be formulated as follows:

- Phase and amplitude freedom outside the reconstruction window g_r: outside the finite region $A\times B$ in the reconstruction plane, called reconstruction window, g_r, the phase and amplitude values for $g(x,y)$ can vary freely. The coding process of a phase-only CGH consists of finding a distribution $g(x',y')$ that satisfies the hologram modulation restrictions and ensures that the desired object $f(x,y)$ is contained within g_r, figure 3.
- Phase freedom inside g_r: for intensity objects, the phase $\varphi(x,y)$ can vary freely inside the region *gr*.
- Freedom of the scale factor α: One can vary the scale factor α in order to obtain an error reduction inside the region g_r, figure 2.

Assuming now a full complex-amplitude modulation CGH, there is much more flexibility in the design, considering there is always at least one solution to the problem. The complex values of transmittance of each hologram's element are directly represented by the CGH distribution $g(x',y')$, with no iterative procedures. Using this approach, the total computer time can be reduced by two orders of magnitude (Neto, et al., 2003, 2004), and the quality of reconstructed images are quite better than a phase-only CGH, as can be see in section 3.3.

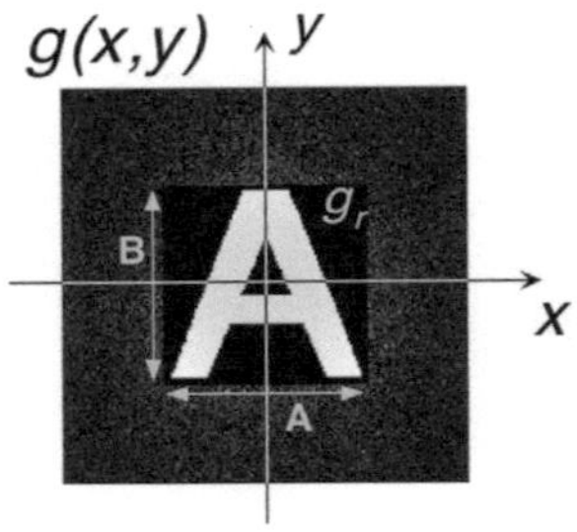

Fig. 3. Reconstruction window used to improve the DOE reconstruction. Outside the region g_r of dimensions *AXB*, the phase and amplitude values for $g(x,y)$ can vary freely. Reprinted with permission of SPIE (Neto, L.G. 2004).

2.3 Diffraction efficiency

The quality of the CGH calculation is evaluated via the degree of deviation of the produced image from the desired one, and via the fraction energy going onto the generation of the image. Each application has specific demands for emphasizing one or another figure of merit. The main three parameters are used (Soifer, et al., 1997):

Diffraction efficiency, η: Is the ratio of the light energy having come into a predefined area (region g_r) to the entire energy striking the DOE:

$$\eta = \frac{\sum_{m=-A/2}^{+(A/2)-1} \sum_{n=-B/2}^{+(B/2)-1} \left| g_{m-Mo,\ n-No} \right|^2}{MN} \tag{5}$$

where *MN* denotes the extension of the hologram when illuminated by a light wave with a constant amplitude equal to unity. It is also helpful to describe η as the product of the transmitted, η_T, and diffracted, η_D, terms: $\eta = \eta_D \eta_T$. The term η_T gives the fraction of the energy transmitted through the hologram and the term η_D gives the fraction of the transmitted energy that forms the signal:

$$\eta_D = \frac{\sum_{m=-A/2}^{+(A/2)-1} \sum_{n=-B/2}^{+(B/2)-1} \left| g_{m-Mo,\ n-No} \right|^2}{\sum_{m=-A/2}^{+(A/2)-1} \sum_{n=-B/2}^{+(B/2)-1} \left| g_{m,n} \right|^2} \qquad (6)$$

and

$$\eta_T = \frac{\sum_{m=-A/2}^{+(A/2)-1} \sum_{n=-B/2}^{+(B/2)-1} \left| G_{k,l} \right|^2}{MN} \qquad (7)$$

where M_0 and N_0 are the center coordinates of the reconstruction window g_r.

For phase-only holograms, where $mod\{G_{mn}\}=1$, one has η_D=1, i.e., all energy is transmitted to form the signal and $\eta=\eta_D$. Under certain circumstances it is possible to achieve diffraction efficiencies near unity.

Considering amplitude-only holograms, it is possible to estimate an upper limit of η, by considering that the hologram absorbs half of the energy of the illuminating wave, resulting $\eta_T = 0.5$. In the image plane, *50%* of the transmitted energy will be concentrated in the zeroth diffraction order, and only *25%* of the transmitted light will form the signal because there is a twin signal. Thus the upper limit of η_D is *25%*. As a result one obtain $\eta \le 12.5\%$. In general η_T is significantly smaller than *50%* for a gray-level hologram, and η may be much lower than the upper limit of *12.5%*. The twin signal in the image plane is present due to the fact that amplitude holograms are real functions, and their Fourier transform must be Hermitian functions (Bracewell, 1986).

Image reconstruction error (root-mean-square deviation), e_i: During the IFTA, at the *i*-th iteraction, is the square root of the sum of the squares of the difference between the produced and desired intensities, with averaging summation performed over the entire domain of definition of image.

$$e_i = \frac{\sum_{m=-A/2}^{+(A/2)-1} \sum_{n=-B/2}^{+(B/2)-1} \left\{ \left| f_{mn} \right| - k_i \left| g^{*i}_{m+Mo,n+No} \right| \right\}^2}{AB}, \quad k_i = \frac{1}{\alpha_i} \qquad (8)$$

$$\alpha_i = \frac{\sum_{m=-A/2}^{+(A/2)-1} \sum_{n=-B/2}^{+(B/2)-1} \left| g^{*i}_{mn} \right|^2}{\sum_{m=-A/2}^{+(A/2)-1} \sum_{n=-B/2}^{+(B/2)-1} \left\{ \left| f_{mn} \right| \times \left| g^{*i}_{m+Mo,n+No} \right| \right\}} \qquad (9)$$

where the symbol (*) denotes the resulting image distribution after hologram constraints being applied in the *i*-th iteration.

Several parameters influence the minimization of the mean square error: the number of iterations, *i*, the pattern of the desired object (signal), the dimension of the reconstruction window, the degrees of freedom and the manner that the extra phase modulation operator is implemented (Wyrowski, 1990).

Signal-to-noise ratio, SNR: Another important parameter to evaluate the quality of the reconstructed signal, is the signal-to-noise ratio (*SNR*), defined by:

$$SNR = \frac{\sum_{m=-A/2}^{+(A/2)-1} \sum_{n=-B/2}^{+(B/2)-1} k_{final} \left| g_{mn}^{final} \right|^2}{AB \times e_{final}} \qquad (10)$$

3. Implementation of computer-generated holograms

This section reports the design and fabrication of several digital holograms, from amplitude-only to full complex-amplitude modulation. In all cases the image generated is intensity-only objects. It also presents the implementation of a diffractive photomask applied to optical lithography and a diffractive lens with engineered Point Spread Function (PSF). The former diffractive element employs optical information processing to implement a spatial filter for target silhouettes applied to passive infrared motion sensors.

3.1 Amplitude-only modulation CGH: Fourier-type CGH, binary intensity-only object

In general it is difficult to implement the amplitude distribution of an amplitude-only CGH. Photographic film has nonlinearities which make difficult the control of the desired amplitude modulation and postscript printers generate amplitude modulation in a binary manner (half toning). Hence one is restricted to a binary amplitude modulation of the light, and the hologram elements can only assume the binary values *0* (when light is blocked) and *1* (when light passes through).

In the case of amplitude holograms one can assume that for the same signal there is a direct correspondence between the distribution of binary amplitude hologram and the distribution of a binary phase hologram (Wyrowski, 1990). Hence, during the coding of the hologram one can consider the binary amplitude hologram as a binary phase transmittance which can only have the complex value $1.exp[-j\pi] = -1$ and $1.exp[-j0] = +1$ ($j = (-1)^{1/2}$). By the addiction of a DC-term, it is possible to transform the resulting binary phase distribution into a binary amplitude distribution with values *0* and *1*. Since the hologram elements present only real values, the light distribution generated is Hermitian (Bracewell, 1986).

Figure 4 shows results of numerical and optical reconstructions of a pattern generated by a *64X64* pixels amplitude-only modulation hologram. The reconstruction plane has extension of *256X256* pixels. The hologram distribution can also be seen. The element was fabricated by using photographic film.

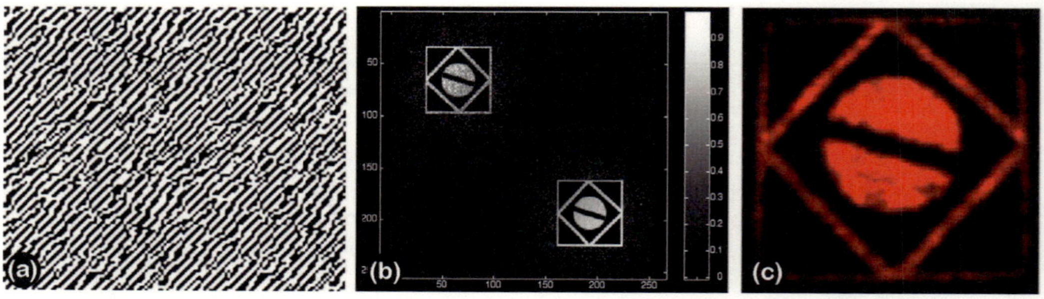

Fig. 4. (a) *64X64* pixels binary amplitude-only modulation hologram; (b) numerical reconstruction, with a reconstruction plane of *256X256* pixels; (c) Optical reconstruction. Only one image is shown; the DC central spot and the twin image were suppressed.

The main limitation in using a photographic film is the low diffraction efficiency achieved by an amplitude element, maximally of *12.5%*, if compared with the diffraction efficiency of about *90%* of a phase element (Wyrowski, 1990).
There are other methods to produce amplitude-only holograms such as Detour Phase Holograms, proposed by (Brown & Lohmann, 1969) and the Direct Binary Search, proposed by (Seldowitz, Allebach & Sweeney, 1987).

3.2 Phase-only modulation CGH: Fourier-type CGH, binary intensity-only object

Because the resulting inverse light propagation distribution $f(x',y')$ usually presents values of phase and amplitude modulation, this distribution must be adapted to the possible DOE distribution $g(x',y')$, which usually is a phase-only, due to its high diffraction efficiency. The use of phase modulation is attractive because no energy is absorbed by the hologram. A 2π phase modulation can be achieved in quantized values (O'Shea et al., 2004), or in continuous values (Turunen & Wyrowski, 1997; Herzig, 1997). If reduced quantized phase values are used, the problem of stagnation can arise during the iterative method. To overcome this limitation, the same IFTA algorithm (figure 2) can be used, with modifications to include the quantized phase values (Wyrowski, 1990).
Considering phase-only CGH, the IFTA method is started by placing the information of the object, f_{mn}, multiplied by random phase values, in the region of reconstructed window g_r. Since the hologram values are not restricted to real values, there is no twin image in the reconstruction plane, and g_r can be placed in its center. It is possible to improve the results for both amplitude- and phase-only CGHs by introducing corrections in the reconstruction window. The first correction, introduced by Wyrowski, takes account of the amplitude error from the previous reconstruction (Wyrowski, 1990a), and the second correction, introduced by Fienup, which includes a phase rotation operator inside g_r that tends to rotate the phase of the new input $g_{mn(i+1)}$ in the opposite direction to the phase of the old input $g_{mn(i)}$ (Fienup, 1980). Figure 5 shows optical reconstructed images, and microrelief profiles from binary phase-only CGH. Typical values for phase-only CGH are e_{final} = 2x10-3, η = 72% and *SNR*= 95.

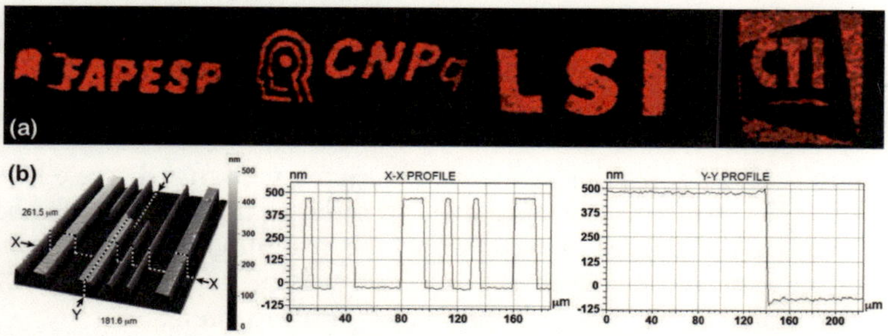

Fig. 5. (a) Optical reconstruction from binary phase-only CGH, (b) microrelief profiles obtained by interferometric microscopy technique. Reprinted with permission of Elsevier (Cirino, G. A. 2010).

3.3 Complex-amplitude modulation CGH: diffractive lithography photomask

The huge importance of the lithography on the fabrication of micro devices is well known. For most applications binary photolithography masks are employed. These masks consist of a transparent plate, covered by a patterned opaque film that acts as a light intensity modulating agent. The phase shift mask technique, introduced in the beginning of the eighties, employs a transparent thin film that modulates the phase of the transmitted light in

order to compensate the diffraction effect of the edges of the mask features (Levenson et al. 1982, Levenson 1993, 1994, 2006). In this application example, a phase shift diffractive proximity printing lithographic mask was designed, manufactured and tested (Cirino et al., 2010b). Its design is based on a Fresnel-type CGH, with phase and amplitude modulation, with a binary intensity-only object to be generated.
The proposed diffractive structure was designed to form the required image within a plane at *50 μm* behind the mask (distande *d* of figure 1b), onto a resist coated silicon wafer, as shown in figure 6. In this figure the mask is located in the plane *(x',y',z=0)*, and the wafer under exposure is located in the plane *(x,y,z=50μm)*. This inter-plane distance may be chosen as most suited for the used alignment equipment. This gives another important degree of freedom for this procedure, optimizing the resolution of the transferred patterns for specific equipment.

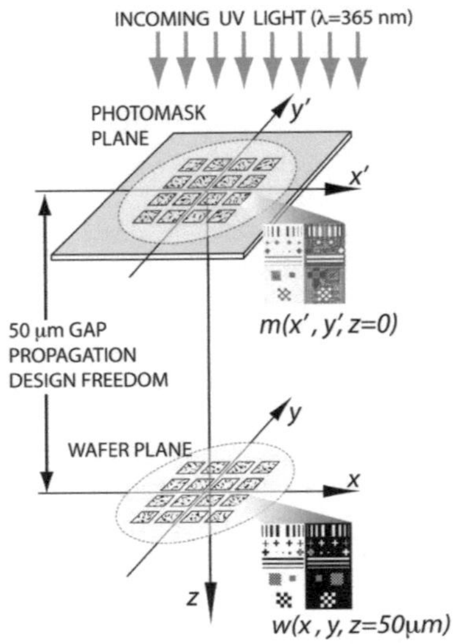

Fig. 6. Schematic view of the optical field projected by the mask, when illuminated by a UV plane wave. The pattern is obtained by applying free space light propagation along the *50 μm* distance. Reprinted with permission of OSA (Cirino, G. A., 2010).

The obtained amplitude and phase distributions were mapped into discrete levels. A coding scheme using sub-cells structure was employed in order to increase the number of discrete levels, thus increasing the degree of freedom in the resulting mask. The results show an improvement of the achieved resolution - linewidth as good as *1.5 μm* - what is impossible to obtain with traditional binary masks in proximity printing mode.
The photomask was fabricated by binary optics (O'Shea et al., 2004) and the sub-cells-based coding scheme was introduced to increase the design degree of freedom. Another contribution of this work is the employment of an a:C-H thin film to modulate the light amplitude. The amorphous carbon layer blocks the light only in the UV region, being partially transparent in the visible region of the spectrum. This characteristic makes mask alignment much easier. A fused silica substrate is used to modulate the phase and the a:C-H layer to modulate the intensity of the incoming UV light plane wave.
In order to give additional degree of freedom and expand the possible values attributed to the complex transmittance of the mask, consider the cells of the photomask shown in figure 7. The cell centers are spaced at regular intervals of width X in the x' direction and regular intervals of width Y in the y' direction. Since each cell is divided in sub-cells, it is possible to deal with

the relation between areas that assume two different values of phase and/or amplitude inside this cell.

Consider, for example, the first quarter of complex plane. Figure 7 illustrates the subdivision of a cell, formed by 4×4 sub-cells, with new possible values of phase and amplitude. Figure 7a shows the structure of a particular cell that can modulate the phase between *0* and *π/2*. Intermediate phase values of the function *ϕ(x',y')*, distributed between *0* and *π/2*, can be approximated by combining different regions inside the cell.

The variation of *V*, from *0* to *Y* inside the cell will introduce a smooth phase variation from *0* to *π/2* radians.

Figure 7b shows the structure of a window created on top of the phase cell of figure 7a, in order to modulate the amplitude of incident light between *0* and *1*. The variation of *U*, from *0* to *X* inside the cell will introduce a smooth amplitude variation from *0* to *1*.

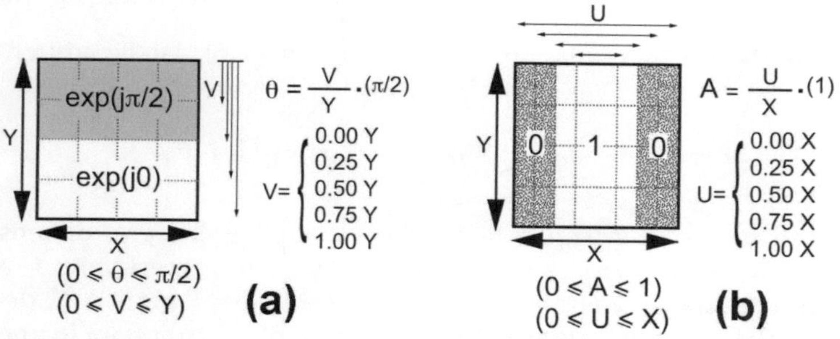

Fig. 7. Subdivision of a cell, formed by 4 × 4 sub-cells. (a) structure of a particular cell that can modulate the phase between *0* and *π/2*, depending of the value of *V*; (b) structure of a window created over the phase cell of figure 7a, in order to modulate the amplitude of incident light between *0* and *1*, depending of the value of *U*. Reprinted with permission of OSA (Cirino, G. A., 2010).

Figure 8a shows the structure of a particular cell with complex transmittance of amplitude equal to *0.5* and phase equal to *π/4*. Figure 8b shows all the possible values (black dots) of amplitude and phase within the first quadrant, as well as, the particular value of the cell of figure 8a. The full complex-amplitude modulation is achieved on the entire complex plane, i.e., on the regions of second, third and fourth quadrant, by combining, respectively, phase values *π/2-π*, *π-3π/2*, and *3π/2-0*.

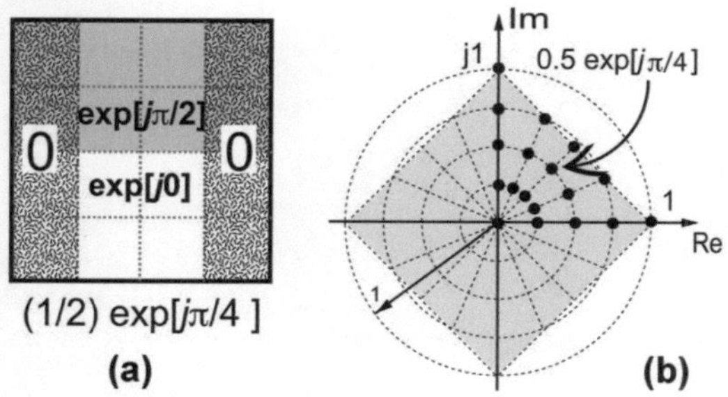

Fig. 8. (a) Structure of a particular cell with complex transmittance of amplitude equal to *0.5* and phase equal to *π/4*; (b) all the possible value (black dots) of amplitude and phase within the first quadrant, as well as, the particular value of the cell of figure (a). Reprinted with permission of OSA (Cirino, G. A., 2010).

To generate the optical reconstruction, the scalar diffraction theory was employed, by considering the free space propagation of the light along the gap between mask and wafer planes. The mask was designed by the inverse propagation of light. In this method, the desired complex distribution of the mask $m(x',y',z=0) = w(x',y',z=0)$ is obtained by solving the Helmholtz equation $(\nabla^2 + k^2)w(x,y,z) = 0$ in the frequency domain for a predefined intensity light distribution $w(x,y,z=50\mu m)$, which must be present in the wafer plane, as shown in figure 6.
The application of the requirement above shows that $W(fx,fy,z=50\mu m)$ must satisfy the differential equation (Goodman, 1996):

$$\frac{\partial^2}{\partial z^2}W\left(f_x\,,f_y\,,z\right)+\left(\frac{2\pi}{\lambda}\right)^2\left[1-\left(\lambda f_x\right)^2-\left(\lambda f_y\right)^2\right]W\left(f_x\,,f_y\,,z\right)=0 \quad (11)$$

An elementary solution of this equation, in the frequency space, can be written in the form:

$$W\left(f_x\,,f_y\,,z\right)=M\left(f_x\,,f_y\,,0\right)\exp\left[j\frac{2\pi}{\lambda}z\sqrt{1-\left(\lambda f_x\right)^2-\left(\lambda f_y\right)^2}\right]\,,\ j=\sqrt{-1} \quad (12)$$

where $W(fx,fy,z=50\mu m)$ and $W(fx,fy,z=0) = M(fx,fy,z=0)$ are the two-dimensional Fourier transform of the functions $w(x,y,z=50\mu m)$ and $m(x',y',z=0)$, respectively; f_x and f_y are the frequency coordinates in the x and y directions, respectively. Equation 12 describes, in the frequency space, the relation between the desired light distribution in the wafer plane $w(x,y,z=50\mu m)$ and the diffraction structure in the mask plane, $m(x',y',z=0)$.
From the above result, the propagation phenomenon may be regarded as a linear, dispersive spatial filter with a finite bandwidth of extension inside a circular region with radius λ^{-1} in the frequency plane. Its transfer function, $H(f_x,f_y)$, is given by

$$H\left(f_x\,,f_y\right)=\begin{cases}\exp\left[j\frac{2\pi}{\lambda}z\sqrt{1-\left(\lambda f_x\right)^2-\left(\lambda f_y\right)^2}\right] & \text{for } \sqrt{\left(f_x\right)^2+\left(f_y\right)^2}<\frac{1}{\lambda}\\ 0 & \text{otherwise}\end{cases} \quad (13)$$

Figure 9 shows a schematic diagram of the phenomenon of free space propagation, and its relation with frequency and image spaces, as well as mask and wafer planes.

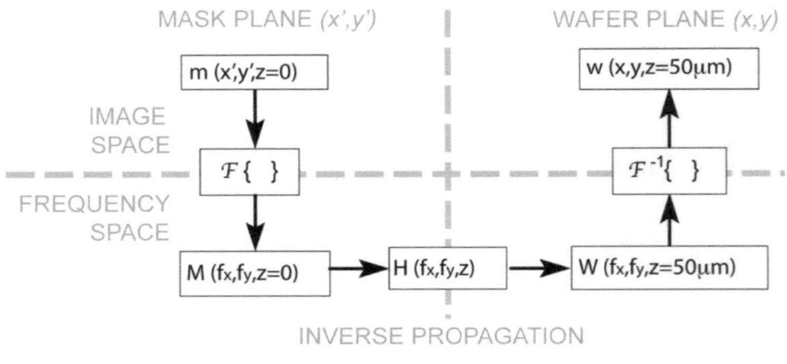

Fig. 9. Schematic diagram of the phenomenon of free space propagation, and its relation with frequency and image spaces, as well as mask and wafer planes. Reprinted with permission of OSA (Cirino, G. A., 2010).

To calculate the inverse propagation from the wafer plane to the mask plane, one has to apply the inverse filter for propagation:

$$m(x',y',z=0)=\Im^{-1}\left\{\frac{W\left(f_x\,,f_y\,,z=50\mu\text{m}\right)}{\exp\left[j\frac{2\pi}{\lambda}z\sqrt{1-\left(\lambda f_x\right)^2-\left(\lambda f_y\right)^2}\right]}\right\} \quad (14)$$

where $\Im\{.\}$ denotes the two-dimensional Fourier transform operator.
In the calculation of the inverse light propagation, the following normalized function, $m_N(x',y')$, is considered,

$$m_N\left(x',y'\right)=\beta\, m\left(x',y'\right)=A\left(x',y'\right)\exp\left[j\phi\left(x',y'\right)\right] \quad (15)$$

where $A(x',y') = |m(x',y')|$ is the amplitude $[0 \leq A(x',y') \leq 1]$, $\phi(x',y')$ is the phase distribution $[0 \leq \phi(x',y') \leq 2\pi]$, $\beta=1/\{MAX\ [m(x',y')]\}$ is a scale factor (a real number) used to normalize the maximum amplitude of $|m(x',y')|$ to unity, and the operator $MAX[.]$ represents the maximum value of the function $m(x',y')$. In the design, the desired reconstruction is a real distribution $w(x,y,z=0) = |w(x,y,z=0)|\ exp[j0]$.
The diffractive phase shift mask is considered as a two-dimensional matrix structure with $M\times N$ rectangular cells of dimensions $X\times Y$, with M and N being even numbers. Each cell represents a complex transmittance value $m_{kl}(x',y')$ with $[(-M/2) \leq k \leq ((M/2)-1)]$ and $[(-N/2) \leq l \leq ((N/2)-1)]$.
The phase and amplitude information must be mapped into the discrete values:

$$A\left(x',y'\right)\exp\left[j\phi\left(x',y'\right)\right] \leftrightarrow A_{kl}\left(x',y'\right)\exp\left[j\theta_{kl}\left(x',y'\right)\right] \quad (16)$$

Considering the cell structure described above and supposing that the mask is placed in the (x',y') plane, the mathematical representation of the diffractive mask in terms of Fourier optics is given by:

$$m_{kl}\left(x',y'\right)=rect\left(\frac{x'}{MX},\frac{y'}{NY}\right)\left\{\sum_{k=-M/2}^{M/2-1}\sum_{l=-N/2}^{N/2-1} rect\left(\frac{x'-kX}{A_{kl}}\right)\left[\begin{array}{l} rect\left(\frac{y'-lY}{Y}\right)+ \\ \left(e^{j\pi/2}-1\right) rect\left(\frac{y'-Y\left(\frac{1}{2}-\frac{\theta_{kl}}{\pi}\right)-lY}{Y-\frac{2\theta_{kl}}{\pi}}\right)\end{array}\right]\right\} \quad (17)$$

Figure 10a shows the intensity distribution of a test structure desired to be transferred to the wafer, $w(x,y,z=50\mu m)$; figures 10b and 10c show the corresponding amplitude distribution, $A(x',y')$ $[0<A<1]$, and phase distribution, $\phi(x',y')$ $[0<\phi<2\pi]$, to be implemented in the mask after applying the propagation method described above.

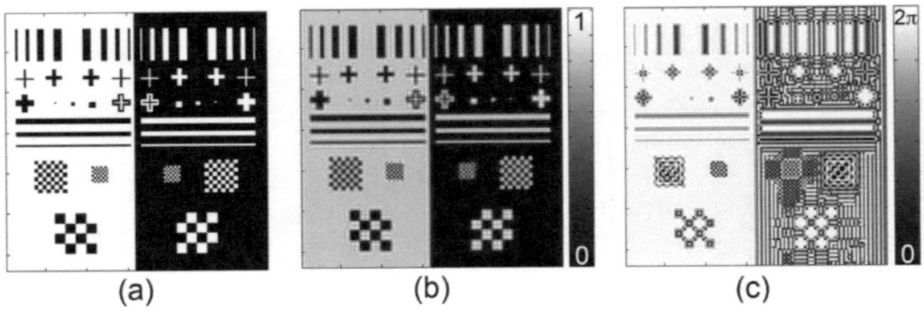

Fig. 10. Lithographic test structures used to show the basic idea of the complex-amplitude modulation proximity printing mask. (a) Numerical reconstructed image of the desired light distribution; (b) amplitude $A(x',y')$ and (c) phase $\phi(x',y')$ distributions of the resulting phase shift photomask. Reprinted with permission of OSA (Cirino, G. A., 2010).

Figure 11 shows schematically the entire process to fabricate the proposed diffractive phase shift proximity photomask.

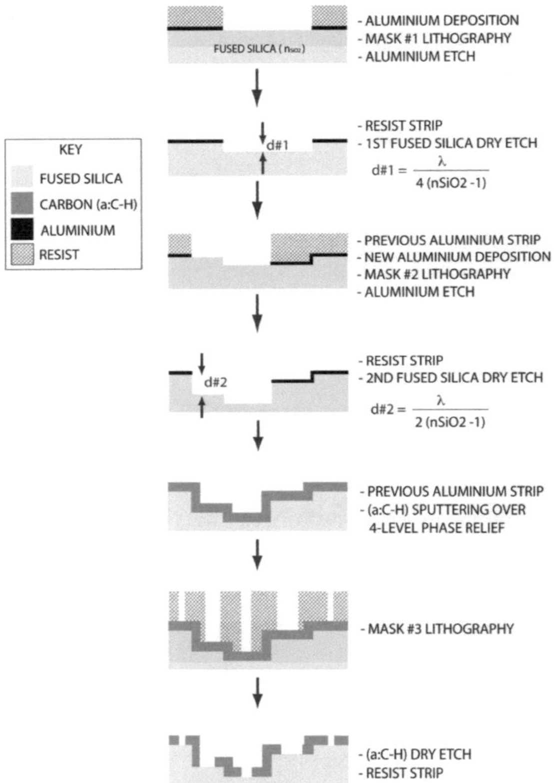

Fig. 11. Schematic view of the entire process to fabricate the proposed diffractive phase shift proximity photomask. Reprinted with permission of OSA (Cirino, G. A., 2010).

The phase shift diffractive photomask was implemented using the discrete phase and amplitude informations. The phase distribution was transferred by generating a variation of the thickness $d_{kl}(x',y')$ on a fused silica substrate with refractive index n_{SiO2}, using the relation

$$d_{kl}(x',y') = \frac{\theta_{kl}(x',y')}{2\pi}\frac{\lambda}{(n_{SiO2}-1)} \tag{18}$$

where $\lambda = 365nm$ is the operating wavelength of the exposure light source.
This variable thickness was approximated by using binary optics (O'Shea et al., 2004), with four phase-levels.
To implement the amplitude modulation in this phase shift diffractive mask, an a:C-H layer is deposited on top of the SiO2 four-level phase relief, and patterned by a third lithographic mask (mask #3 in figure 11), generating the variable apertures size over the region of each phase pixel structure previously patterned. It should be mentioned that this carbon layer is opaque at $\lambda = 365\ nm$, and relatively transparent in the visible region of the spectrum.
Figure 12 shows part of the test structures with a detail of a line 1.5 μm wide by *82 μm* length, patterned in proximity (*50 μm* gap) and contact modes. One can note that the line obtained by proximity exposure is clearly resolved while that obtained by contact exposure is barely resolved.

(a) 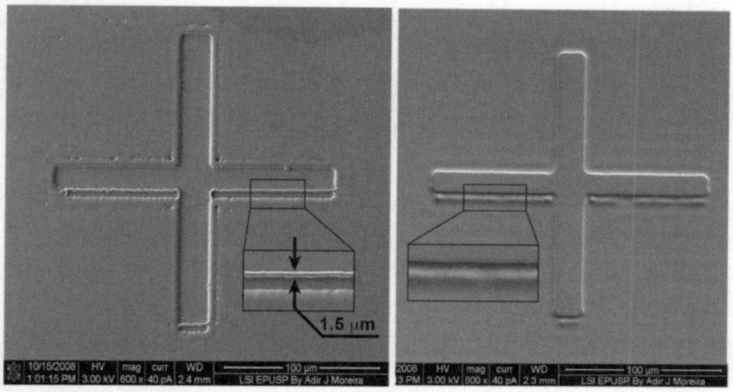 (b)

Fig. 12. SEM micrographs of part of the test structures with a detail of a line *1.5 μm* wide by *82 μm* length, patterned in (a) proximity (*50 μm* gap) and (b) contact modes. The line obtained by proximity exposure is clearly resolved while that obtained by contact exposure is barely resolved. Reprinted with permission of OSA (Cirino, G. A., 2010).

3.4 Complex-amplitude modulation CGH: Fresnel-type CGH, gray-tone intensity-only object

In this section a Fresnel-type CGH with a graytone intensity-only object was designed and fabricated (Neto et al., 2003,2004). The calculations procedure are the same as previous section. A hologram was designed considering the reconstruction plane located of *1.2* meters from the hologram (distance *d* of figure 1b). Here again, the desired reconstruction was a real distribution $f(x,y,0) = |f(x,y,0)|.\ exp\ (j0)$ with smooth variations intensity (gray-tone diffraction pattern). The hologram pixel size is *40 μm* by *40 μm* and the total hologram size is *1024 μm* by *1024 μm*.
In figure 13 are shown the desired reconstruction distribution *f(x,y,d)* together with its corresponding phase and amplitude distributions, obtained after applying the propagation methods described in the previous section.
The complete device fabrication sequence is shown schematically in figure 14a. All the used process steps are well controlled and well known for micromechanical and microelectronic applications. The fabrication is based on DLC thin film, with four phase level surface relief. After the reactive ion etching, a reflective aluminum layer is deposited over the phase relief

using a thermal evaporation process. The amplitude modulation is achieved by removing small parts of this reflective aluminum layer over the region of each phase pixel. The area of these apertures must be proportional to the amplitude $|G_{kl}|$., following the sampling scheme described in equations 1-4.

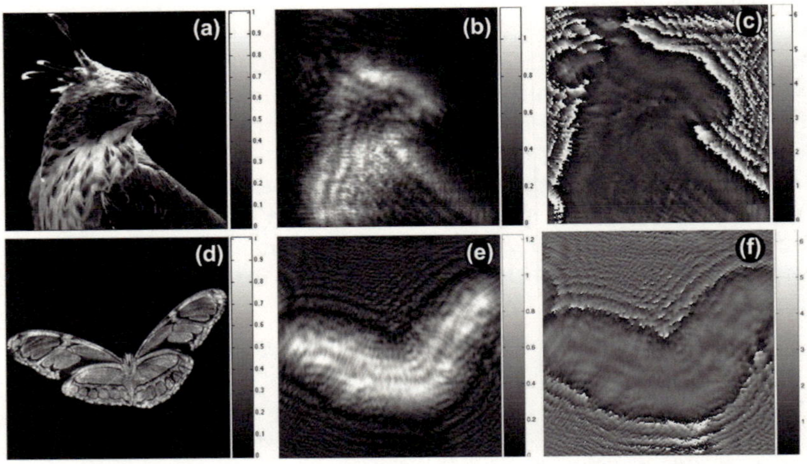

Fig. 13. (a&d) Image of the desired light wave front distribution $f(x,y,z) = = |f(x,y)| exp[i\varphi(x,y)]$, (b&e) amplitude ($0 \leq |f(x,y)| \leq 1$) and (c&f) phase distribution of the resulting Fresnel hologram, ($0 \leq \varphi(x,y) \leq 2\pi$). Reprinted with permission of SPIE (Neto, L. G., 2004).

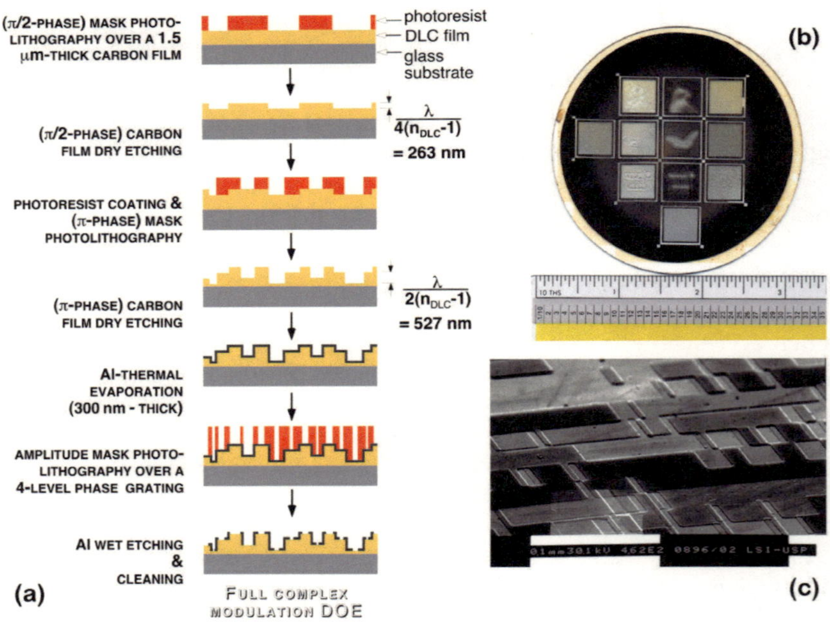

Fig. 14. (a) Schematic view of the fabrication of the full complex amplitude modulation CGH; (b) 3-inch diameter glass substrate, with 12 different CGHs fabricated. (c) Scanning electron microscope picture with details of the geometry of the rectangular pixels. Reprinted with permission of SPIE (Neto, L. G., 2004).

A photolithographic mask with apertures over the region of each phase pixel was fabricated considering the amplitude distribution $|G_{kl}|$.
Each value of $|G_{kl}|$, $(0 \leq |G_{kl}| \leq 1)$, is coded in a 8×8 cell, positioned over each respective pixel of phase. With the used mask set, 11 different DOEs onto a 3-inch diameter glass substrate were fabricated. A photography of a 3 inch wafer with the fabricated elements is shown in Fig. 14b, and a micrograph with details of the geometry of a rectangular pixel that forms the hologram structure, is shown in Fig. 14bc
Figure 15 compares the desired reconstruction *f(x,y,d)* of Fig. 13a with the optical reconstruction of the Fresnel hologram fabricated using the amplitude and phase information of Figs. 13b and 13c. Because a constant phase distribution is assumed in the reconstruction plane (the desired reconstruction is a real distribution), no speckle noise is present in the optical reconstruction.

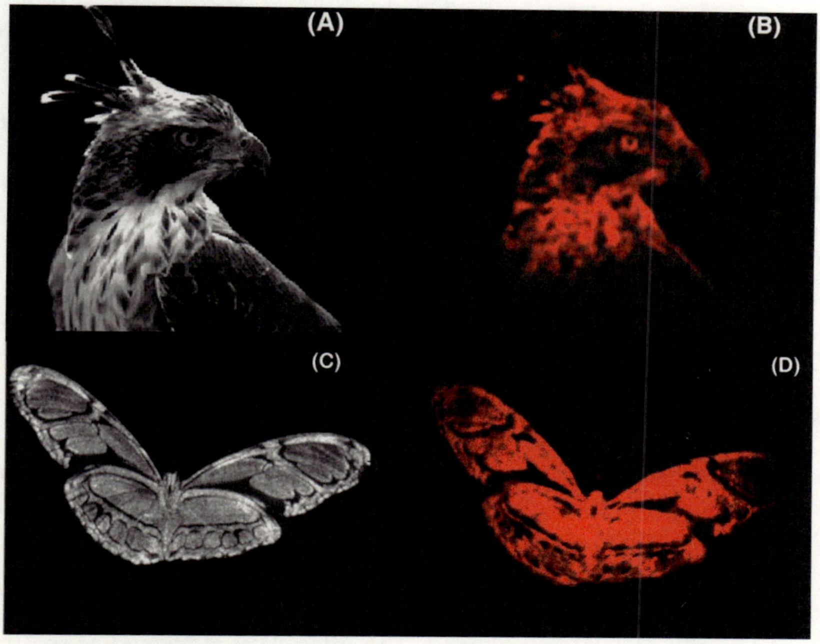

Fig. 15. Comparison between the desired and the obtained reconstructions of a gray-tone object. (a&c) Numerical distribution *f(x,y,d)*, (b&d) optical reconstruction of the Fresnel CGH fabricated using the amplitude and phase information of Figs. 13b and 13c. Reprinted with permission of SPIE (Neto, L. G., 2004).

For comparison purpose, figure 16a shows the phase distribution of a phase-only CGH designed using one hundred iterations of the IFTA, for the same desired image shown in Fig. 13a. The reconstruction plane was also located *1.2* meters away from the element. A random phase distribution was attributed to in the first iteration. The phase-only CGH was manufactured using a similar process sequence described above (of course, without the process steps concerning the third mask which determines the aluminum features).
Figure 16b shows the computer reconstruction of the phase-only element and Fig. 16c shows its optical reconstruction: a strong speckle noise is present in the optical reconstruction.

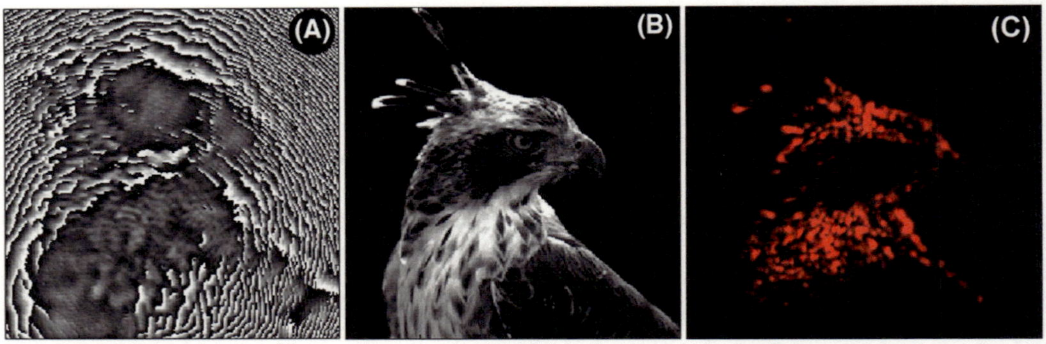

Fig. 16. Results obtained by a phase-only CGH for the same desired reconstruction of figure 13 (a) phase distribution of the CGH, designed using one hundred iterations of the IFTA for the desired image shown in Fig. 13a, (b) computer reconstruction, (c) optical reconstruction: a strong speckle noise is present. Reprinted with permission of SPIE (Neto, L. G., 2004).

The comparison between figures 15b and 16c indicate clearly the superiority of the new element over the traditional phase only DOE. The improved optical performance is caused by the total freedom in the design assured by the full complex amplitude modulation. Naturally, the diffraction efficiency of figure 16b is lower than that of figure 17c, due to the amplitude modulation mask (figure 13b), which blocks a significant portion of the incoming wavefront.

3.5 Complex-amplitude modulation CGH: Fourier-type CGH, binary intensity-only object

In this example a binary, intensity-only image is generated, considering nine levels of amplitude modulation and four levels of phase modulation as a constraint in the hologram plane. The reconstruction window g_r (signal window) in the reconstruction plane has the same space-bandwidth product of the hologram, i.e., one has no complex amplitude freedom in the reconstruction plane. In this plane one has only phase freedom assuming the object is a intensity-only distribution. After 100 iterations, the resulting $128X128$ complex amplitude hologram $G_{kl} = |G_{kl}|\ exp(j\phi_{kl})$,with $(0 \leq |G_{kl}| \leq 1)$, $\phi_{kl} = (0, \pi/2, 2\pi/2, 3\pi/2)$ and $(1 \leq k,l \leq 128)$, is obtained. The diffraction efficiency of the transmitted energy, considering equation 7, is $\eta_T \cong 22\%$. The quality of the resulting complex amplitude hologram G_{kl} is measured using the mean square error function, defined by equation 8, which compares the final hologram's reconstruction g_{mn} with the desired reconstruction f_{mn} is $e_{final} = 2.1\times 10^{-3}$.

The resulting phase distribution was used in the fabrication, here again, based on sputtered carbon thin film, with four phase level surface relief, as described in the previous section. Figure 17a shows the complex-amplitude transmittance of the resulting hologram. Here again, it presents a sub-cell approach design, which enable an increase in the degree of freedom in the coding of complex distributions in the hologram media, resulting in high quality images (diffraction patterns). Figures 17b and 17c shows the intensity of the computer reconstruction compared to the intensity of its optical reconstruction.

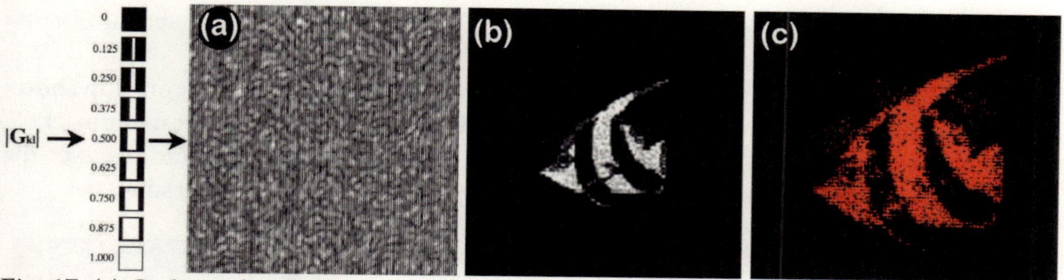

Fig. 17. (a) Coding of each value of $|G_{kl}|$, $(0 \leq |G_{kl}| \leq 1)$ in a 8×8 cell, positioned over each respective pixel of phase, (b-c) comparison between (left) the intensity of the hologram computer reconstruction g_{mn} and (right) the intensity of its optical reconstruction.

3.6 Optical information processing: Silhouette filtering by a diffractive lens

Passive InfraRed (PIR) motion sensors have been extensively used in electronic monitoring of assets in domestic and corporative security applications. Its principle of operation is based on the employment of a pyroelectric detector, which is responsible for transducing an incident IR radiation (*8-14 μm* wavelength range) to an electric signal.

In this application, the engineered Point Spread Function (PSF) concept of optical systems using cubic phase distribution was applied, in order to implement an optical image filtering in the design of inexpensive Fresnel lenses array for PIR motion sensors. The goal is to perform, in a purely optical approach, the so-called pet immunity feature, i.e., the sensor must detect humans, filtering and neglecting the information generated by small pets (Cirino, 2003, 2004).

Imaging systems that use incoherent illumination obey the intensity convolution integral (Goodman, 1996):

$$I_{out}(x,y) = \int_{-\infty}^{\infty}\int \left| h(x-u, y-v) \right|^2 I_G(u,v)\, du dv \tag{19}$$

where $I_G(u,y)$ is the intensity of the image predicted by geometrical optics for each lens of the lenses array, $h(x,y,u,v)$ is the PSF, or the impulse response of the lens and $I_{Out}(x,y)$ is the intensity of the output image formed by the lens. u and v are the coordinates in the spatial frequency space in the x- and y-directions, respectively. Thus in this general case, one can regard the output image as being a convolution of the image predicted by geometrical optics with the intensity of the PSF of the system. Narrow PSF results in optical system with the ability to resolve higher spatial frequencies compared to a system with a wide PSF.

Considering only the vertical y-direction, human silhouettes have lower spatial frequencies compared to small pets. In order to assure the distinction, a one-dimensional vertical cubic phase distribution $P_Y(y)$ is superimposed to the conventional quadratic phase distribution of each lens. $P_Y(y)$ is defined by (Dowski, 1995):

$$P_Y(y) = \exp(j\sigma\, y^3) \tag{20}$$

where σ is the cubic phase factor. The multiplication of the spherical phase distribution of each lens by the cubic phase $P_Y(y)$ spreads the resulting PSF. Considering equation 19, the convolution of $I_G(u,y)$ with the intensity of a wide PSF performs a low-pass spatial filtering in the y-direction.

Figure 18a shows the input test image used in the computer simulation. Figure 18b shows the resulting distribution using equation 19, and figure 18c shows its corresponding PSF. One can note from fig.3b that, thanks to this particular PSF distribution, the image of the human body has a higher intensity distribution compared to the image of the small pet.

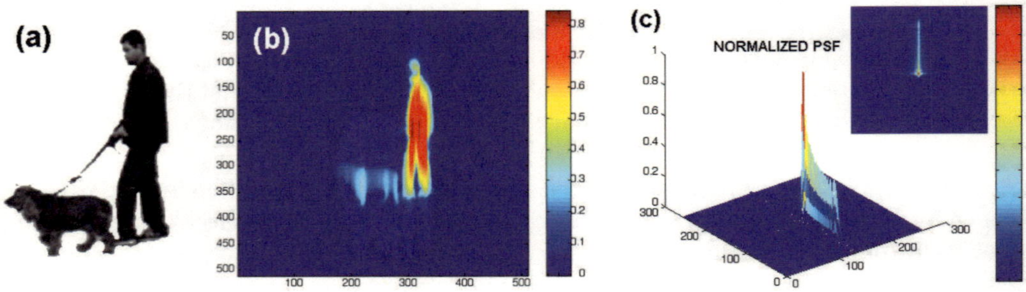

Fig. 18. Numerical simulation of the performance of the cubic phase mask to be embedded in the sensor's Fresnel lens array (a) Input test image used in the computer simulation; (b) corresponding simulated output image, normalized to its maximum value; (c) normalized PSF distribution of the optical system. One can note that the image of the human silhouette has higher brightness when compared to the image of a small pet.

In order to show how the lens can be employed in the distinction of human targets from pet targets, a lens set on Diamond-Like Carbon (DLC) thin film were designed and manufactured by microfabrication process techniques (Cirino, 2004; Goodman 1996). Although the sensor operates at the *8-14* μm wavelength range, a diffractive Fresnel lens which operates in the visible portion of the spectrum was fabricated, with a central wavelength of *500 nm*. This is a suitable way to show that the principle works, being directly applied in the desired wavelength range (mid infra-red). For comparative purposes, besides each cubic-phase distribution Fresnel lens, a conventional quadratic Fresnel lens was also fabricated. Both lenses were submitted to the same process steps during the whole fabrication process. The lenses microrelief were fabricated on the *1.5* μm-thick DLC thin film, using the following relation

$$th(x_P, y_P) = \frac{\lambda}{2\pi(n_{DLC} - 1)} \phi(x_P, y_P) \tag{21}$$

where $th(x_p, y_p)$ is the thickness variation within the transparent carbon film, $\lambda = 500\,nm$ is the operating wavelength, n_{DLC} is the refractive index of the *DLC* layer (measured at wavelength λ) and $\phi(x_p, y_p)$ is the phase distribution information $[0 \le \phi(x_p, y_p) \le 2\pi]$ (Herzig, 1997).

Figure 19 shows schematically the experimental set up employed to perform the comparative tests. The object, located in the back focal plane of the lens, at a distance z_i, is illuminated by a diffused incoherent white light source. The lens produces the image of the object directly in the plane of a CCD camera, located at a distance z_a from the lens.

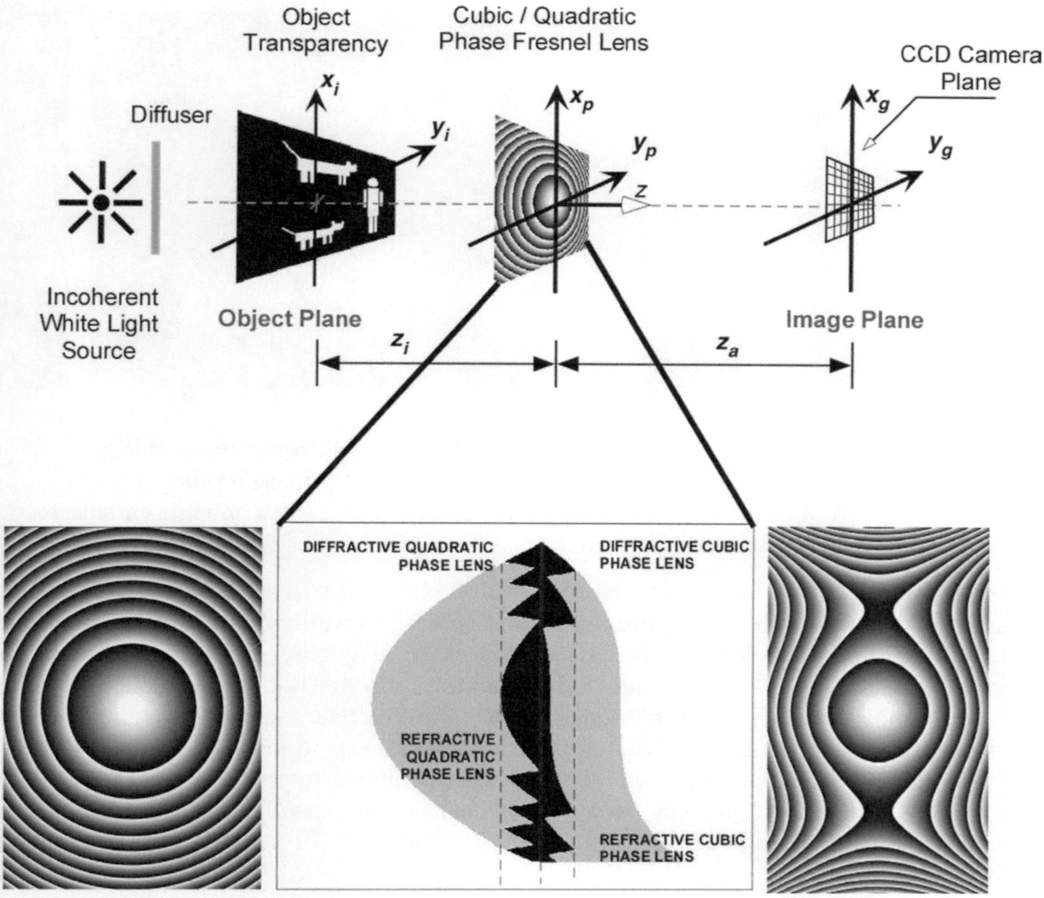

Fig. 19. Experimental set up employed to store the images produced by the fabricated Fresnel lenses. The object plane is illuminated by a diffused incoherent white light source and the lens images the object directly in the CCD plane of the camera. Details of the quadratic and cubic phase distributions are also shown.

Figure 20a and 20b show the resulting images from the conventional lens (quadratic phase profile) and the proposed lens (cubic phase profile), respectively. One can see that cubic phase profile spreads the light as predicted by the simulated results. The optical power generated by pet targets is attenuated (to a value of about 50% of the maximum value), as compared with human targets. Therefore this approach can be used to allow some blurring of the PSF of the optical system in such a way that it turns possible the discrimination between humans and pets for passive infrared motion sensor applications.

4. Conclusions

In this chapter some aspects of digital holography were delineated. An iterative method based on the fast Fourier transform was presented, which permits the design of diffractive optical elements in the scalar domain.

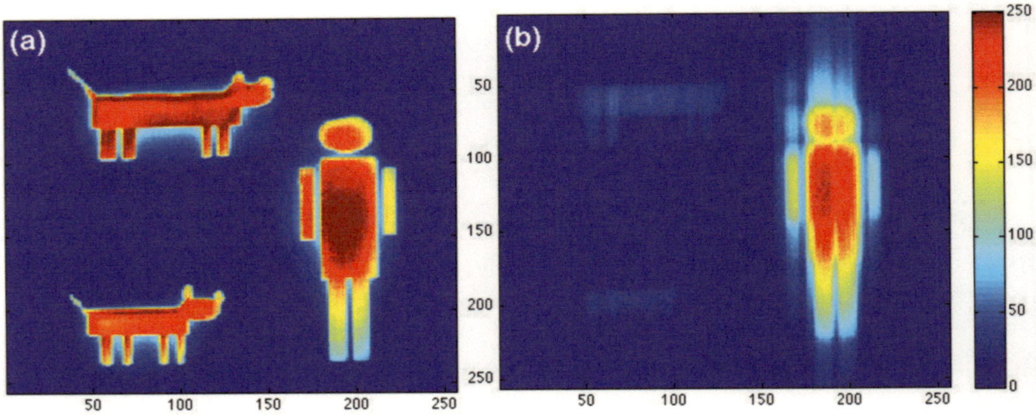

Fig. 20. Resulting intensity images generated from the Fresnel lenses in the visible range of spectrum. (a) Resulting image from the conventional (quadratic phase profile) lens; (b) resulting image from the proposed lens (cubic phase profile). One can note the enhancement of the human body intensity distribution

Several examples of Fresnel- and Fraunhofer-type holograms as well as amplitude-only, phase-only and full complex-amplitude modulation devices were presented to illustrate the potential of this technology in the generation of high quality diffraction patterns. A fabrication technique based on a patented carbon thin film process was employed. Among the applications, a diffractive photomask and a diffractive lens with a cubic phase distribution were presented. The first application enables one to increase the resolution of features to be transfer to a silicon wafer when operating in proximity printing mode lithography. The second one was used to implement a low-pass filter in order to make a distinction of human silhouettes from those of small pets for passive infrared motion sensors applications.

5. Acknowledgment

The authors would like to thank CTI for the fabrication of the photolithographic masks, and the agencies Conselho Nacional de Desenvolvimento Cientfico e Tecnológico (CNPq), process # 151908/2007-9, and Fundação de Amparo à Pesquisa no Estado de São Paulo (FAPESP), grants # 98/02755-7 ; 00/10181-2 ; 00/11117-6 and 02/07840-0. This work is part of the Instituto Nacional de Ciência e Tecnologia: Óptica e Fotônica (INCT-INOF), sponsored by CNPq (grant # 573587/2008-6) and FAPESP (grant # 08/57858-9).

6. References

Benton, S. A. (1969) On a method for reducing the information content of holograms, *Journal of Optical Society of America*, Vol. 59, pp. 1545-1551, ISSN 0030-3941

Bracewell, R. N., (1986), *The Fourier Transform and Its Applications*, McGraw Hill Series in Electrical Engineering, ISBN 0-0730-3938-1, Massachusetts, USA

Brown, B.R. & Lohmann, A.W. (1966) Complex spatial filtering with binary masks, *Applied Optics* (June 1966), v.5, n.6, p.967-969, ISSN 2155-3165

Brown, B. R.; Lohmann, A. W. (1969a) Computer-generated binary holograms, *IBM Journal of Research and Development*, v.13, pp. 160-168, ISSN 0018-8646

Brown, B.R. & Lohmann, A.W. (1969b)"Computer-generated binary holograms", *IBM Journal of Research and Development* (May 1969), v.13, p. 160-168, ISSN 0018-8646

Cirino, G. A. & Neto, L. G. (2003), Design of cubic distribution lenses for passive infrared motion sensors, *Proceedings SPIE 5073 of the XXV Thermosense*, pp. 476–484, ISBN 0-8194-4932-6, Orlando, Florida, USA, April 22-24, 2003

Cirino, G. A. & Neto, L. G. (2004), Optical implementation of cubic-phase distribution lenses for passive infrared motion sensors, *Proceedings SPIE 5405 of the XXVI Thermosense*, pp. 189–198, ISBN 0-8194-5328-5, Orlando, Florida, USA, April 13-15, 2004

Cirino, G. A.; R. D. Mansano; P. Verdonck, L. Cescato, & L. G. Neto (2010a), Diffractive phase-shift lithography photomask operating in proximity printing mode, *Optics Express* (August 2010), Vol. 18, Issue 16, pp. 16387-16405, ISSN 1094-4087

Cirino, G. A.; R. D. Mansano; P. Verdonck, R. G. Jasinevicius, & L. G. Neto (2010b), Diffraction gratings fabricated in DLC thin films, *Surface & Coatings Technology* (June 2010), Vol. 204, No. 18-19, pp. 16387-16405, ISSN 0257-8972

Cooley, J. W. & Tukey, J. W. (1965) An Algorithm for the Machine Calculation of Complex Fourier Series, *Mathematics of Computation*, vol. 19, No. 90, pp. 297–301, ISSN 1088-684

Denisyuk, Y. N. (1962) . Photographic reconstruction of the optical properties of an object in its own scattered radiation field. *Soviety Physics - Doklady*, Vol. 7, pp. 543-549, ISSN 0038-5689

Dowski Jr., E. R. & Cathey, W. T. (1995), Extended depth of field through wave-front coding, *Applied Optics,* Vol. 34 No. 11, pp. 1859-1866, ISSN 2155-3165

Fienup, J. R. (1980), Iterative method applied to image reconstruction and computer-generated holograms, *Optical Engineering* (June 1985), Vol. 19, No. 3, pp. 297-305, ISSN 0091-3286

Gabor, D (1948) A new microscopic principle, *Nature* (1948), Vol. 161, pp. 777-778, ISSN 0028-0836

Goodman, J. W. (1996), *Introduction to Fourier Optics*, pp. 55-61, McGraw Hill, ISBN 0-07-024254-2, New York, USA

Herzig, H. P. (1997). *Micro-optics: Elements, Systems and Applications,* Taylor & Francis, ISBN 0-7484-0481-3, London, England

Leith, E.N. & Upatnieks, J. (1962), Wavefront reconstruction and communication theory, *Journal of Optical Society of America* (October 1962), Vol. 52, Issue 10, pp. 1123-1128, ISSN 1520-8532

Levenson, M. D.; Viswanathan, N. S. & Simpson, R. A. (1982), Improving resolution in photolithography with a phase-shifting mask, *IEEE Transactions on Electron Devices* (December 1982), ED-29, Issue 12, pp. 1828-1836, ISSN 0018-9383

Levenson, M. D. (1993), Wavefront engineering for photolithography, *Physics Today* (July 1993) pp. 28-36, ISSN 0031-9228

Levenson, M. D. (1994), Extending the lifetime of optical lithography technologies with wavefront engineering, *Japanese Journal of Applied Physics* (September 1994), Vol. 33, pp. 6765-6773, ISSN 1347-4065

Levenson, M. D. (2006), Using destructive optical interference in semiconductor lithography, *Optics & Photonics News* (April 2006), pp. 30-35, ISSN 1047-6938

Lohmann, A. W. (1956), Optische Einseitenbandübertragung Angewandt auf das Gabor-Mikroskop," *Optica Acta* (June 1956), Vol. 3, pp. 97-100, ISSN 0030-3909

Lohmann, A.W. & Paris, D.P. (1967) Binary Fraunhofer holograms, generated by computer, *Applied Optics* (October 1967), Vol. 6, No.10, pp.1739-1748, ISSN 2155-3165

Lohmann, A. W. (2008). A Pre-History of Computer-Generated Holography, *Optics and Photonics News* (February 2008), pp. 36-41, ISSN 1047-6938

Mait, J. N. (1995) Understanding diffractive optic design in the scalar domain, Journal of Optical Society of America A (October 1995), Vol. 12, No. 10, pp. 2145-2158, ISSN 084-7529

Neto, L. G.; Roberto, L. B.; Mansano, R. D.; Verdonck, P.; Cirino, G. A. & Steffani (2001), M. A. Multiple Line Generation Over High Angle Using Hybrid Parabolic Profile and Binary Surface-Relief Phase Element, *Applied Optics* (January 2001), vol. 40, issue 2, pp.211-218, ISSN 2155-3165

Neto, L. G.; Cardona, P. S. P.; Cirino, G. A.; Mansano, R. D. & Verdonck, P. (2003) Design, fabrication, and characterization of a full complex-amplitude modulation diffractive optical element, *Journal of Microlithography, Microfabrication and Microsystems*, (April 2003), vol. 2, issue 2, pp. 96-104, ISSN 1537-1646

Neto, L. G.; Cardona, P. S. P.; Cirino, G. A.; Mansano, R. D. & Verdonck, P. (2004), Implementation of Fresnel full complex-amplitude digital holograms, *Optical Engineering*, (November 2004), vol. 43, No. 11, ISSN 0091-3286

Neto, L. G.; Mansano, R. D.; Cirino, G. A.; Zambom, L. S. & Verdonck, P. (2008), Amorphous hidrogenated carbon film (June 2008), *US Patent* 7,381,452

O'Shea, D. C.; Suleski, T. J.; Kathman, A. D. & Prather, D. W. (2004), *Diffractive Optics: Design, Fabrication, and Test*, Tutorial Texts in Optical Engineering Vol. TT62, SPIE Press, ISBN 0-8194-5171-1, Washington, USA

Seldowitz, M. A.; Allebach, J. P. & Sweeney, D. W., 1987, Sinthesis of digital holograms by direct binary search, *Applied Optics*, Vol. 26, No. 14, pp. 2788-2797, ISSN 2155-3165

Turunen, J.; Wyrowski, F. (1997). *Diffractive Optics for Industrial and Commercial Applications*, Akademie Verlag, ISBN 3-05-501733-1, Berlin, Germany

Wyrowski, F. (1989) Iterative quantization of digital amplitude holograms, *Applied Optics* (September 1989), Vol. 28, No. 18, pp. 3864-3870, ISSN 2155-3165

Wyrowski, F. (1990a) Diffractive optical elements: iterative calculation of quantized, blazed phase structures, *Journal of Optical Society of America A* (June 1990), vol. 7, pp. 961-969, ISSN 084-7529

Wyrowski, F. (1990b) Diffraction efficiency of analog and quantized digital amplitude holograms: analysis and manipulation, *Journal of Optical Society of America A* (March 1990), vol. 7, pp. 383-389, ISSN 084-7529

3

Electron Holography of Magnetic Materials

Takeshi Kasama[1],
Rafal E. Dunin-Borkowski[2] and Marco Beleggia[1]
[1]*Technical University of Denmark,*
[2]*Forschungszentrum Jülich,*
[1]*Denmark*
[2]*Germany*

1. Introduction

Transmission electron microscopy (TEM) involves the use of high-energy (60-3000 keV) electrons that have passed through a thin specimen to record images, diffraction patterns or spectroscopic information from a region of interest. Many different TEM techniques have been developed over the years into highly sophisticated methodologies that have found widespread application across scientific disciplines. Because the TEM has an unparalleled ability to provide structural and chemical information over a range of length scales down to atomic dimensions, it has developed into an indispensable tool for scientists who are interested in understanding the properties of nanostructured materials and in manipulating their behavior (Smith, 2007).

State-of-the-art TEMs are now equipped with spherical and chromatic aberration correctors and can provide interpretable image resolutions of 0.05 nm (Erni et al., 2009). However, in addition to conventional TEM techniques that can be used to provide structural and compositional information about materials, the TEM also allows magnetic and electrostatic fields in specimens to be imaged with nanometer spatial resolution. One of the most powerful techniques for providing this information is electron holography, which was originally proposed as a means to compensate for lens aberrations and to improve electron microscope resolution (Gabor, 1949). Electron holography is still the only technique that provides direct access to the phase shift of the electron wave that has passed through a thin specimen, in contrast to more conventional TEM techniques that record only spatial distributions of image intensity.

Electron holography has only recently become widely available on commercial electron microscopes. The earliest studies using electron holography were restricted by the limited brightness and coherence of the tungsten filaments that were used as electron sources (Haine & Mulvey, 1952). The availability of high brightness, stable, coherent field emission electron guns now allows electron holography to be applied to a wide variety of materials such as quantum well structures, magnetic thin films, semiconductor devices, natural rocks and biominerals.

Although there are several forms of electron holography, including in-line and scanning TEM-based electron holography (Cowley, 1992), the most commonly-used form is the TEM mode of off-axis electron holography, which involves using an electrostatic biprism that is inserted into the microscope column perpendicular to the electron beam (Möllenstedt & Düker, 1956). As described in detail below, the application of a voltage to the biprism results in overlap of the electron wave to create an interference fringe pattern. Other forms of electron holography are described elsewhere (e.g., Cowley & Spence, 1998).

Off-axis electron holography can be divided further into two modes: (i) High-resolution electron holography, in which the interpretable resolution in a high-resolution TEM image can be improved by the use of phase plates to correct for microscope lens aberrations (Lichte, 1991; Lehmann et al., 1999) and (ii) Medium-resolution electron holography, in which magnetic and electrostatic fields in materials can be studied quantitatively, usually with nanometer spatial resolution (Tonomura, 1992; Dunin-Borkowski et al., 2004). Some of the most successful early examples of the application of medium-resolution electron holography include the experimental confirmation of magnetic flux quantization in superconducting toroids (i.e., the Aharonov-Bohm effect) and the study of magnetic flux vortices in superconductors (Tonomura et al., 1986; Matsuda et al, 1989). Here we concentrate on more recent applications of medium-resolution electron holography for the study of magnetic materials.

This chapter begins with an outline of the basis and requirements for electron holography and the procedures that are typically used to obtain amplitude and phase information from electron holograms, including microscope calibrations for magnetic studies and the practical steps that are required to extract the magnetic contribution to the signal from a reconstructed phase image. We then provide selected examples of the application of electron holography to the characterization of magnetic domain structures in nanostructured materials. We describe a novel approach for measuring the magnetic moment of a chosen nanocrystal quantitatively, before discussing future opportunities for the development and application of electron holography for the study of magnetic materials and devices.

2. Basis of off-axis electron holography

Off-axis electron holography in the TEM involves the examination of a thin specimen using a highly coherent field emission electron source. A schematic diagram showing the typical electron-optical configuration is shown in Fig. 1a. The region of interest is positioned so that it covers approximately half the field of view. A biprism (a thin conducting wire), which is usually located close to the first image plane, has a positive voltage applied to it in order to overlap a "reference" electron wave that has passed through vacuum with the electron wave that has passed through the specimen. The overlap region contains interference fringes and an image of the specimen, in addition to Fresnel fringes originating from the edge of the biprism wire (Figs 1b-e). The electron-optical configuration is equivalent to the use of two electron sources S_1 and S_2, as shown in Fig. 1a (Dunin-Borkowski et al., 2004). By applying a larger voltage to the biprism wire, the separation of the two virtual sources and the width of the overlap region W increase (Figs 1b-d), placing a greater constraint on the spatial coherence of the source required to maintain sufficient interference fringe contrast.

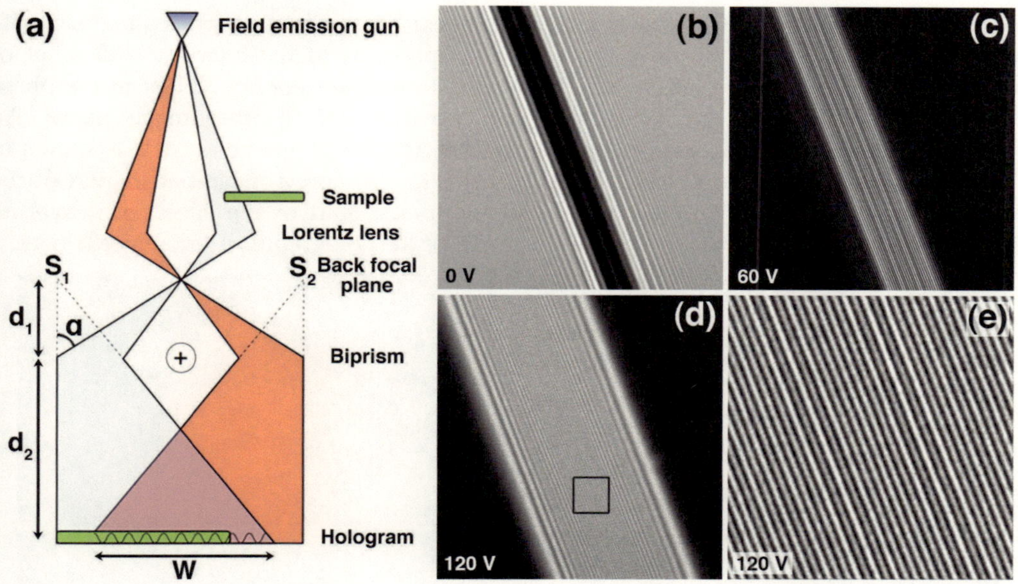

Fig. 1. (a) Schematic diagram of the setup for off-axis electron holography. The symbols are defined in the text. (b-e) Interference fringe patterns (electron holograms) recorded using biprism voltages of (b) 0 V, (c) 60 V and (d,e) 120 V. In (b), only Fresnel fringes from the edges of the biprism wire are visible. Image (e) corresponds to the box marked in (d). The field of view in (b)-(d) is 916 nm.

2.1 Intensity distribution in bright-field TEM images and electron holograms

In coherent image formation in the TEM, the electron wavefunction in the image plane can be written in the form

$$\psi_i(\mathbf{r}) = A_i(\mathbf{r})\exp[i\phi_i(\mathbf{r})] \tag{1}$$

where $\mathbf{r}$ is a two-dimensional vector in the plane of the sample, and A and ϕ refer to amplitude and phase, respectively. In a conventional bright-field TEM image, the intensity distribution imaged on a recording medium is given by the expression

$$I(\mathbf{r}) = |A_i(\mathbf{r})|^2 \tag{2}$$

and all information that relates directly to the phase is lost. In contrast, the intensity distribution in an off-axis electron hologram can be obtained by considering the addition of a tilted plane reference wave to the complex specimen wave in the form

$$I_{hol}(\mathbf{r}) = |\psi_i(\mathbf{r}) + \exp[2\pi i \mathrm{q_c} \bullet \mathbf{r}]|^2 \tag{3}$$

where the tilt of the reference wave is specified by the two-dimensional reciprocal space vector $\mathbf{q} = \mathbf{q}_c$. By combining this expression with Eq. 1, it can be rewritten in the form

$$I_{hol}(\mathbf{r}) = 1 + A_i^2(\mathbf{r}) + 2A_i(\mathbf{r})\cos[2\pi i \mathrm{q_c i} \bullet \mathbf{r} + \phi_i(\mathbf{r})]\,, \tag{4}$$

highlighting the fact that the intensity in an electron hologram consists of three separate contributions: the reference image intensity, the specimen image intensity and a set of cosinusoidal fringes, whose local phase shifts and amplitudes are equivalent to the phase and amplitude, respectively, of the electron wavefunction in the image plane. An experimental electron hologram of a specimen that contains magnetic crystals is shown in Fig. 2a, alongside a vacuum reference hologram (Fig. 2b) and a magnified region of the specimen hologram (Fig. 2c). The amplitude and phase shift of the specimen wave are recorded in the intensity and position, respectively, of the holographic interference fringes.

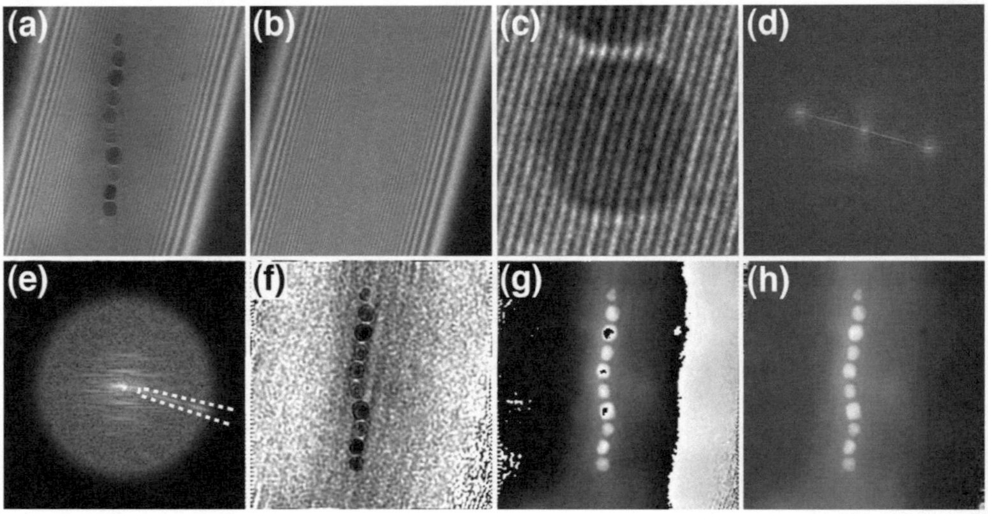

Fig. 2. Sequence of processing steps required to convert a recorded electron hologram into an amplitude and phase image. (a) Experimentally acquired hologram of a region of interest (a chain of magnetite crystals in a magnetotactic bacterium). Coarser fringes, resulting from Fresnel diffraction at the edges of the biprism wire, are visible on either side of the overlap region. The overlap width and holographic interference fringe spacing are 650 and 3.3 nm, respectively. The field of view is 725 nm. (b) Vacuum reference hologram acquired immediately after the specimen hologram. (c) Magnified region of the specimen hologram, showing changes in the positions of interference fringes in a particle. (d) Fourier transform of the electron hologram shown in (a), containing a centerband, two sidebands and a diagonal streak resulting from the presence of Fresnel fringes in the hologram. (e) One of the sidebands extracted from the Fourier transform, shown after applying a circular mask with smooth edges to reduce its intensity radially to zero. If required, the streak from the Fresnel fringes can be removed in a similar manner, by assigning a value of zero to pixels inside the region shown by the dashed line. Inverse Fourier transformation of the sideband is used to provide a complex image wave, which is then displayed in the form of (f) an amplitude image and (g) a modulo 2π phase image. Phase unwrapping algorithms are used to remove the 2π phase discontinuities from (g) to yield the final unwrapped phase image shown in (h).

2.2 Reconstruction of electron holograms

In order to extract phase and amplitude information from an electron hologram, it is usually first Fourier transformed. Based on Eq. 4, the Fourier transform of an electron hologram can be written in the form

$$FT[I_{hol}(\mathbf{r})] = \delta(\mathbf{q}) + FT[A_i^2(\mathbf{r})] + \delta(\mathbf{q}+\mathbf{q_c}) \otimes FT[A_i(\mathbf{r})\exp[i\phi_i(\mathbf{r})]]$$
$$+\delta(\mathbf{q}-\mathbf{q_c}) \otimes FT[A_i(\mathbf{r})\exp[-i\phi_i(\mathbf{r})]] \tag{5}$$

Equation 5 describes a peak at the origin of reciprocal space corresponding to the Fourier transform of the reference image, a second peak centered at the origin corresponding to the Fourier transform of a conventional bright-field TEM image of the sample, a peak centered at $\mathbf{q} = -\mathbf{q}_c$ corresponding to the Fourier transform of the desired image wavefunction and a peak centered at $\mathbf{q} = +\mathbf{q}_c$ corresponding to the Fourier transform of the complex conjugate of the wavefunction. Figure 2d shows the Fourier transform of the hologram shown in Fig. 2a. In order to recover the complex electron wavefunction, one of the two "sidebands" in the Fourier transform is selected (Fig. 2e) and then inverse Fourier transformed. The amplitude and phase of the complex image wave are calculated by using the expressions

$$A = \sqrt{\mathrm{Re}^2 + \mathrm{Im}^2} \tag{6}$$

$$\phi = \tan^{-1}\left(\frac{\mathrm{Im}}{\mathrm{Re}}\right) \tag{7}$$

where Re and Im are the real and imaginary parts of the complex image wavefunction, respectively. Figures 2f and 2g show the resulting amplitude and phase images, respectively. The amplitude image is similar to an energy-filtered bright-field TEM image since the contribution of inelastic scattering to holographic interference fringe formation is negligible (Verbeeck et al., 2011). The phase image is initially calculated modulo 2π, meaning that 2π phase discontinuities appear at positions where the phase shift exceeds this amount (Fig. 2g). If required, the phase image can then be "unwrapped" by using suitable algorithms (e.g., Takeda et al., 1982), as shown in Fig. 2h.

In Fig. 2d, the streak from the "centerband" toward the sidebands is attributed to the presence of Fresnel fringes from the biprism wire, which have a range of spacings. The contribution of this streak to the sideband can lead to artifacts in reconstructed amplitude and phase images and can be minimized by masking it from the Fourier transformtion before inverse Fourier transformation (Fig. 2e). Alternatively, the Fresnel fringes can be eliminated by introducing a second biprism close to a different image plane in the microscope column (Yamamoto et al., 2004; Harada et al., 2005).

As phase information is stored in the lateral displacements of holographic interference fringes, long-range phase modulations arising from inhomogeneities in the charge and thickness of the biprism wire, as well as from lens distortions and charging effects (e.g., at apertures), can introduce artifacts into the reconstructed wavefunction. In order to take these effects into account, a reference hologram is usually obtained from vacuum alone by removing the specimen from the field of view without changing the electron-optical parameters of the microscope (Fig. 2b). Correction is then possible by performing a complex division of the specimen wavefunction by the vacuum wavefunction in real space and calculating the phase of the resulting complex wavefunction to obtain the distortion-free phase of the image wave.

2.3 Magnetic and mean inner potential contributions to the phase shift

The phase shift recorded using electron holography is sensitive to both the electrostatic potential and the in-plane component of the magnetic induction in the specimen. Neglecting dynamical diffraction (i.e., assuming that the specimen is thin and weakly diffracting), the phase shift can be written in the form

$$\phi(x,y) = \phi_e(x,y) + \phi_m(x,y) = C_E \int_{-\infty}^{+\infty} V(x,y,z)dz - \frac{e}{\hbar} \int_{-\infty}^{+\infty} A_z(x,y,z)dz \tag{8}$$

where V is the electrostatic potential, A_z is the component of the magnetic vector potential parallel to the electron beam direction and

$$C_E = \left(\frac{2\pi}{\lambda}\right)\left(\frac{E+E_0}{E(E+2E_0)}\right) \tag{9}$$

is a constant that depends on the microscope accelerating voltage $U=E/e$. In Eq. 9, λ is the (relativistic) electron wavelength and E_0=511 keV is the rest mass energy of the electron. C_E is a constant that depends on the microscope accelerating voltage and takes values of 1.01×10^7, 8.64×10^6 and 6.53×10^6 rad/Vm at accelerating voltages of 80, 120 and 300 kV, respectively. If no external charge distributions or applied electric fields are present within or around the specimen, then the only electrostatic contribution to the phase shift originates from the mean inner potential V_0 of the material coupled with thickness variations of the specimen (most importantly at its edge):

$$\phi_e(x,y) = C_E V_0 t(x,y) \tag{10}$$

where $t(x,y)$ is the specimen thickness. If the specimen has uniform composition, then the electrostatic contribution to the phase shift is proportional to the local specimen thickness at position (x,y). Alternatively, should the thickness be known independently, ϕ_e can be used to provide a measurement of the local mean inner potential.

The magnetic contribution to the phase shift carries information about the magnetic flux. The difference between its value at two arbitrary points in a phase image at coordinates (x_1,y_1) and (x_2,y_2)

$$\Delta\phi_m = \phi_m(x_1,y_1) - \phi_m(x_2,y_2) = -\frac{e}{\hbar} \int_{-\infty}^{+\infty} A_z(x_1,y_1,z)dz + \frac{e}{\hbar} \int_{-\infty}^{+\infty} A_z(x_2,y_2,z)dz \tag{11}$$

can be written in the form of a loop integral

$$\Delta\phi_m = -\frac{e}{\hbar} \oint \mathrm{A} \cdot d\mathrm{l} \tag{12}$$

for a rectangular loop formed by two parallel electron trajectories crossing the sample at coordinates (x_1,y_1) and (x_2,y_2) and joined, at infinity, by segments perpendicular to the trajectories. By virtue of Stokes' theorem,

$$\Delta\phi_m = \frac{e}{\hbar} \iint \mathbf{B} \cdot \hat{\mathbf{n}}\, dS = \frac{\pi}{\phi_0} \Phi(S) \tag{13}$$

where $\phi_0 = h/2e = 2.07\times10^{-15}$ Tm^2 is a flux quantum. The phase difference between any two points in a phase image is therefore a measure of the magnetic flux through the whole region of space bounded by two electron trajectories crossing the sample at the positions of these two points. A graphical representation of the magnetic flux distribution throughout the sample can therefore be obtained by adding contours to a recorded phase image. A phase difference of 2π in such an image corresponds to an enclosed magnetic flux of 4.14×10^{-15} Tm^2. The relationship between the magnetic contribution to the phase shift and the magnetic induction can be established by considering the gradient of ϕ_m, which takes the form

$$\vec{\nabla}\phi_m(x,y) = \frac{e}{\hbar}\left[B_y^p(x,y), -B_x^p(x,y)\right] \tag{14}$$

where

$$B_j^p(x,y) = \int_{-\infty}^{+\infty} B_j(x,y,z)dz \tag{15}$$

are the components of the magnetic induction perpendicular to the incident electron beam direction projected in the beam direction. In the special case where (i) stray fields surrounding the sample can be neglected, (ii) the sample has a constant thickness and (iii) the magnetic induction does not vary with z within the specimen, Eq. 14 can be rewritten in the simplified form

$$\vec{\nabla}\phi_m(x,y) = \frac{et}{\hbar}\left[B_y(x,y), -B_x(x,y)\right], \tag{16}$$

in which there is direct proportionality between the magnetic contribution to the phase gradient and the magnetic induction. Equation 16 has a limited validity, however, and must be used with caution if the aim is to quantify the magnetic field strength and its direction on a local scale. For visualization purposes, it can be useful to relate the spacing and direction of phase contours to the direction of the projected magnetic field within the field of view.

It should be noted that the separation of electrostatic and magnetic contributions to the phase shift is almost always mandatory in order to obtain quantitative magnetic information from a phase image. The few instances when this extra step may be avoided include the special case of magnetic domains in a thin film of constant thickness (far from the specimen edge) and a measurement scheme for quantifying magnetic moments that is described in section 4 below.

2.4 Phase shift of an isolated magnetic nanoparticle

Analytical expressions for the phase shift of an isolated spherical magnetic particle, based on Eq. 8, can be derived for a uniformly magnetized sphere of radius a, magnetic induction $B_\perp$ (along y) and mean inner potential V_0 in the form (de Graef et al., 1999)

$$\phi(x,y)\Big|_{x^2+y^2\le a^2} = 2C_E V_0\sqrt{a^2-(x^2+y^2)} + \frac{2}{3}\left(\frac{e}{\hbar}\right)B_\perp a^3\left(\frac{x}{x^2+y^2}\right)\left\{1-\left[1-\left(\frac{x^2+y^2}{a^2}\right)\right]^{\frac{3}{2}}\right\} \tag{17}$$

$$\phi(x,y)\Big|_{x^2+y^2>a^2} = \frac{2}{3}\left(\frac{e}{\hbar}\right)B_{\perp}a^3\left(\frac{x}{x^2+y^2}\right). \tag{18}$$

Graphical representations of Eqs 17 and 18 are shown in Fig. 4 below for a uniformly-magnetized 50-nm-diameter spherical particle of cobalt, on the assumption that V_0 =17.8 V and $B_{\perp}$=1.8 T. The total phase shift is the sum of mean inner potential and magnetic contributions. For a line profile across the center of the particle in a direction perpendicular to $B_{\perp}$, the expressions reduce to

$$\phi(x)\Big|_{|x|\le a} = 2C_E V_0\sqrt{a^2-x^2} + \frac{2}{3}\left(\frac{e}{\hbar}\right)B_{\perp}\left(\frac{a^3-(a^2-x^2)^{\frac{3}{2}}}{x}\right) \tag{19}$$

$$\phi(x)\Big|_{|x|>a} = \frac{2}{3}\left(\frac{e}{\hbar}\right)B_{\perp}\left(\frac{a^3}{x}\right) \tag{20}$$

which can be used for measurements of the in-plane magnetic induction and mean inner potential of the particle by least-squares fitting (an example is shown in section 3.1.1.1). The magnitude of the difference between the maximum and minimum values of the magnetic contribution to the phase shift across a uniformly magnetized sphere of radius a can be determined from Eqs 19 and 20 to be

$$|\Delta\phi_m| = 2.044\left(\frac{e}{\hbar}\right)|B_{\perp}|a^2. \tag{21}$$

For a uniformly-magnetized infinite cylinder of radius a, the equivalent expression is

$$|\Delta\phi_m| = \pi\left(\frac{e}{\hbar}\right)|B_{\perp}|a^2. \tag{22}$$

2.5 Requirements for off-axis electron holography

In order to provide a highly coherent electron beam, a field emission gun electron source is essential. Although no electron source is perfectly coherent, either spatially or temporally, the degree of coherence must be such that an interference fringe pattern of sufficient quality can be recorded within a reasonable acquisition time, during which specimen and/or beam drift must be negligible. It is common practice to adjust the condenser lens astigmatism to make the beam illumination elongated in the direction perpendicular to the biprism wire. Coherence is then maximized in the elongation direction, while the electron flux is maximized in the region of interest. The elongation direction of the illumination has to be aligned at exactly 90° to the biprism wire, as a slight misalignment can lead to a dramatic decrease in interference fringe contrast. In order to maximize fringe visibility, a small spot size, small condenser aperture size, parallel illumination and a low extraction voltage are usually required. In practice, a balance between coherence, intensity and acquisition time must be achieved.

When characterizing magnetic materials, the microscope objective lens is usually turned off because it creates a large (>2 T) magnetic field at the position of the specimen. Instead, a non-immersion Lorentz lens is used as the primary imaging lens. The Lorentz lens is located within the lower pole piece of the objective lens and allows specimens to be imaged at relatively high magnifications in magnetic-field-free conditions. Although the spherical aberration coefficient C_S of the Lorentz lens is large (e.g., 8400 mm for an FEI Titan microscope (Phatak et al., 2008)), resulting in a point resolution of ~2 nm, magnetic fields in most specimens of interest are relatively slowly varying and the spatial resolution for magnetic characterization using electron holography is often limited primarily by a combination of interference fringe spacing and signal to noise considerations.

The electron biprism, which is normally either a Au-coated glass fiber or a Pt wire with a diameter of ~1 μm, is usually positioned in place of one of the selected-area apertures, close to the first image plane. The optimal position for the biprism in the microscope column has been discussed by Lichte (1996). It should be noted that the biprism cannot be located exactly in an intermediate image plane (Smith & McCartney, 1998). Hence, a readjustment of the electron optics is required in most TEMs. For example, the first intermediate lens may be over-excited and the objective lens (or Lorentz lens) then refocused to place the image below the biprism plane (Smith & McCartney, 1998). Lateral movement and rotation of the biprism is a useful feature, especially for studying magnetic materials.

The most common medium used to record electron holograms is a charge-coupled device (CCD) camera, which provides linear output over a large dynamic range and a high detection quantum efficiency. Digital recording is also convenient for subsequent computer image processing, including reconstruction of holograms. A phase sensitivity of $2\pi/100$ can be achieved by the use of digital recording and processing (de Ruijter & Weiss, 1993), in place of the use of an optical bench as a Mach-Zender interferometer (Tonomura, 1992).

2.6 Imaging parameters

Three coupled experimental parameters affect the quality of final reconstructed phase images significantly: (i) overlap width, (ii) interference fringe spacing and (iii) interference fringe visibility. In general, the application of a higher biprism voltage results in a larger overlap width, a finer interference fringe spacing and a decrease in fringe contrast.

2.6.1 Overlap width

The application of a voltage to a biprism wire results in an overlap width that is given by the expression

$$W = 2\left(\frac{d_1 + d_2}{d_1}\right)\left(\alpha\frac{d_1 d_2}{d_1 + d_2} - R\right) \tag{23}$$

where d_1 is the distance between the focal plane and the biprism, d_2 is the distance between the biprism and the image plane, R is the radius of the biprism wire and α is the deflection angle, as defined in Fig. 1a. A thinner biprism wire can result in a larger overlap width for a given interference fringe spacing and can have fewer Fresnel fringes from the edge of the wire. The overlap width varies linearly with biprism voltage, as shown in Figs 3a and d for an FEI Titan TEM operated at 300 kV, and typically takes a value of between 0.5 and 1.5 μm when using the Lorentz lens as the primary imaging lens of the microscope. The overlap width and the interference fringe spacing can be changed by adjusting the strength of the projector lenses after the biprism either with or without slightly exciting the objective lens (Frost et al., 1996; Sickmann et al., 2011).

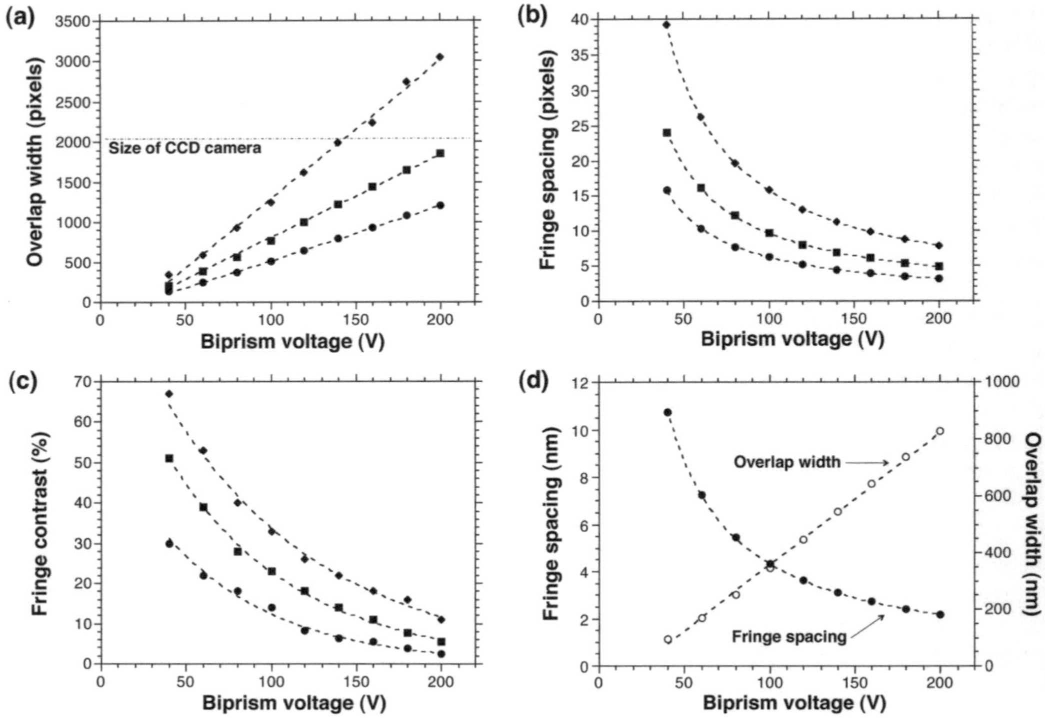

Fig. 3. Measured dependence of (a) overlap width, (b) interference fringe spacing and (c) interference fringe visibility recorded at indicated magnifications of 1450x (circles), 2250x (squares) and 3700x (diamonds) on an FEI Titan 80-300ST TEM operated at 300 kV in Lorentz mode. A Gatan model 894 2K (2048×2048 pixel) UltraScan 1000FT CCD camera mounted in a Gatan imaging filter was used. (d) Interference fringe spacing (filled circles) and overlap width (open circles) in nm plotted as a function of biprism voltage.

2.6.2 Interference fringe spacing

The interference fringe spacing is inversely proportional to the biprism voltage and is given by the expression

$$s = \lambda\left(\frac{d_1 + d_2}{2\alpha d_1}\right), \tag{24}$$

as shown in Fig. 3b for an FEI Titan TEM. Fringe spacings of between 2 and 5 nm are typically used for examining magnetic materials, with a spatial resolution in the final phase image that is presently, at best, approximately three times the measured fringe spacing (Völkl & Lichte, 1990).

2.6.3 Interference fringe visibility

The fringe visibility is one of the most important parameters for electron holography, as it can determine the phase resolution (Lichte, 2008), which is given by the expression

$$\phi_{\min} = \frac{SNR}{\mu}\sqrt{\frac{2}{N_{el}}} \quad (25)$$

where the fringe visibility

$$\mu = \left(\frac{I_{\max} - I_{\min}}{I_{\max} + I_{\min}}\right), \quad (26)$$

SNR is the signal-to-noise ratio in the hologram, N_{el} is the number of electrons collected per pixel and I_{max} and I_{min} are the maximum and minimum intensities of the interference fringes, respectively (Völkl et al., 1995). Figure 3c shows measurements of fringe visibility plotted as a function of biprism voltage for an FEI Titan TEM operated at different magnifications. The difference between the graphs is attributed to the modulation transfer function of the CCD camera, suggesting that the use of a higher magnification can result in better phase resolution, although the field of view then becomes smaller. The recommended number of pixels per interference fringe on a CCD camera is generally chosen to be at least four (Smith & McCartney, 1998). The number of electrons per pixel in a recorded electron hologram is usually at least 100-500 for an acquisition time of 2-8 sec. If the biprism wire and the specimen are stable enough, then a longer acquisition time can be used (Cooper et al., 2007). Furthermore, if two or more biprisms are used simultaneously, then there is more flexibility to control the overlap width, interference fringe spacing and biprism rotation angle independently, as well as to eliminate Fresnel fringes from the recorded image wave (Harada et al., 2006).

2.7 Separation of magnetic and mean inner potential contributions to the phase

Although measurements of the local value of the mean inner potential can be used to provide useful information about the morphology or chemical composition of a specimen (Dunin-Borkowski et al., 2004), its contribution to the phase shift is generally detrimental for studies of magnetic materials using electron holography. In particular, it can be much larger than the magnetic contribution to the phase shift for small (sub-50-nm) magnetic nanocrystals. Figure 4 shows simulations of the mean inner potential and magnetic contributions to the phase shift for a 50-nm-diameter Co crystal that is uniformly magnetized. Although the particle is in a single domain state, the mean inner potential dominates the recorded phase image. Separation of the desired magnetic contribution from the total phase shift is then essential.

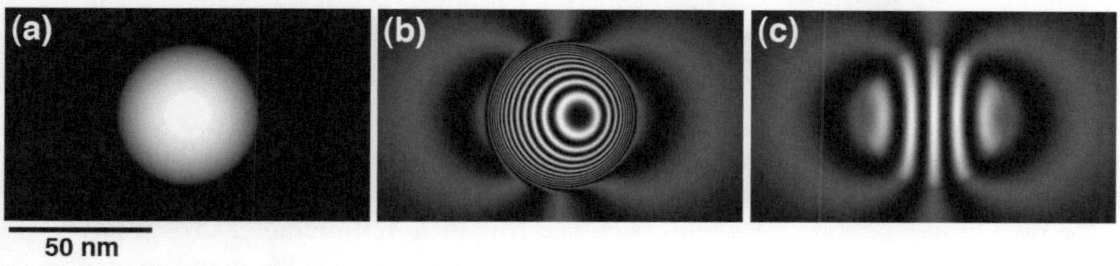

Fig. 4. Simulations of the phase shift of a 50-nm-diameter Co (B=1.8 T; V_0=17.8 V) spherical particle, which is uniformly magnetized in the vertical direction. (a) Thickness map. (b) Total phase shift. (c) Magnetic contribution to the phase shift alone. In (b) and (c), the cosine of 12 times the phase is shown.

In principle, the most accurate way of achieving this separation involves turning the specimen over after acquiring a hologram and acquiring a second hologram from the same region, as shown in Fig. 5. The two holograms are aligned after flipping one of them over digitally (Fig. 5a). Then, their sum and difference are used to determine twice the mean inner potential and twice the magnetic contribution to the phase shift, respectively. The mean inner potential contribution can be subtracted from all subsequent phase images acquired from the same region (Fig. 5b), followed by adding phase contours and colors to form final magnetic induction maps. A degree of smoothing of the final phase images is often used to remove statistical noise and artifacts resulting from misalignment of the pairs of phase images.

An alternative method, which is often more practical, involves performing a magnetization reversal experiment *in situ* in the microscope and then selecting pairs of holograms that differ only in the magnetization direction in the specimen. The addition of two phase images, in which the specimen is oppositely magnetized, provides twice the mean inner potential contribution to the phase. Magnetization reversal can be performed by using either a magnetizing TEM specimen holder or the vertical magnetic field of the conventional microscope objective lens and tilting the specimen. In the latter case, the objective lens can be turned off and the specimen is tilted back to the horizontal before acquiring each hologram. This approach is only applicable if the magnetization in the specimen reverses perfectly. If this is not the case, then reversal measurements may need to be repeated multiple times so that non-systematic differences between reversed images are averaged out.

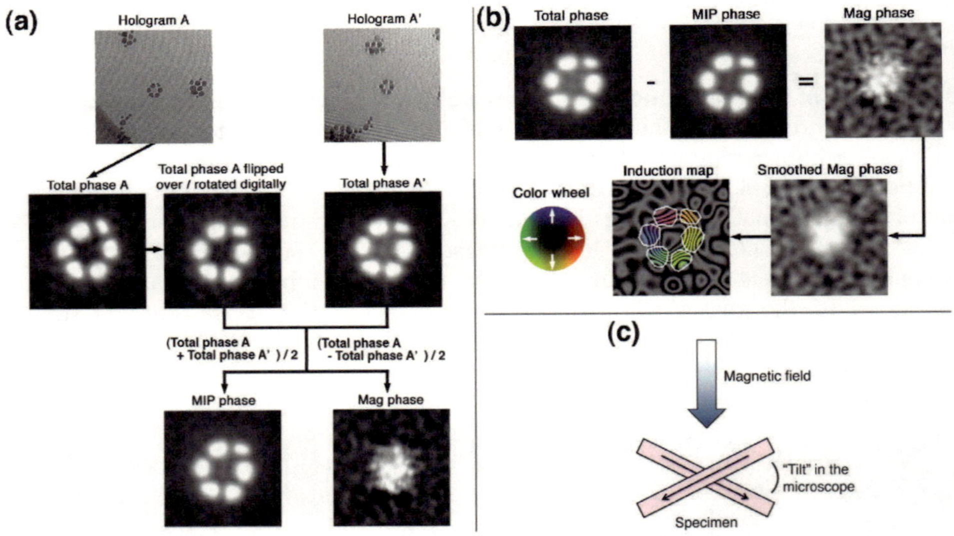

Fig. 5. Experimental procedure used to obtain (a) the mean inner potential (MIP) contribution to the phase shift and (b) the magnetic (Mag) contribution to the phase shift and the final induction map for a self-assembled ring of 25-nm-diameter Co crystals (Kasama et al., 2008). The induction map in (b) shows the cosine of 96 times the magnetic contribution to the phase. The color wheel shows the direction of the projected induction in each crystal. (c) Schematic diagram showing the use of *in situ* tilting of a specimen in the vertical magnetic field of the partially excited microscope objective lens to perform a magnetization reversal experiment. Adapted from Kasama et al. (2007).

If the magnetization in the specimen does not reverse exactly and if the specimen cannot be turned over during an experiment, then it may be possible to generate an "artificial" mean inner potential map either from a t/λ map (where t is the specimen thickness and λ is the inelastic mean free path) (Egerton, 1996) or from a high-angle annular dark-field image. The constant of proportionality between the thickness map and the mean inner potential contribution to the phase can be determined by least-squares fitting to data collected near the specimen edge, where the magnetic contribution to the phase shift is smallest (Harrison et al., 2002). This method may not be applicable in the presence of strong diffraction contrast or if the specimen contains several materials with different compositions. In this situation, the use of two different microscope accelerating voltages or a phase image acquired above the Curie or Néel temperature of the specimen (Loudon et al., 2002) may provide an alternative approach.

2.8 Calibration of the magnetic field of the objective lens

The objective lens of the microscope can generate a magnetic field of up to ~2 T in the direction of the electron beam, which would saturate the magnetization in most samples in this direction. Hence, for magnetic studies, the objective lens is usually either turned off or excited slightly to apply a chosen magnetic field to the sample. Figure 6 shows a Hall probe calibration of the vertical magnetic field at the specimen position in an FEI Titan TEM operated at 300 kV, plotted as a function of normalized objective lens current. Additional lenses such as the mini-condenser and Lorentz lens also affect the magnetic field at the position of the specimen slightly. In practice, however, since the mini-condenser lens setting is fixed and the Lorentz lens is adjusted only slightly for focusing, the resulting change in the magnetic field at the specimen can be as small as ±1 Oe.

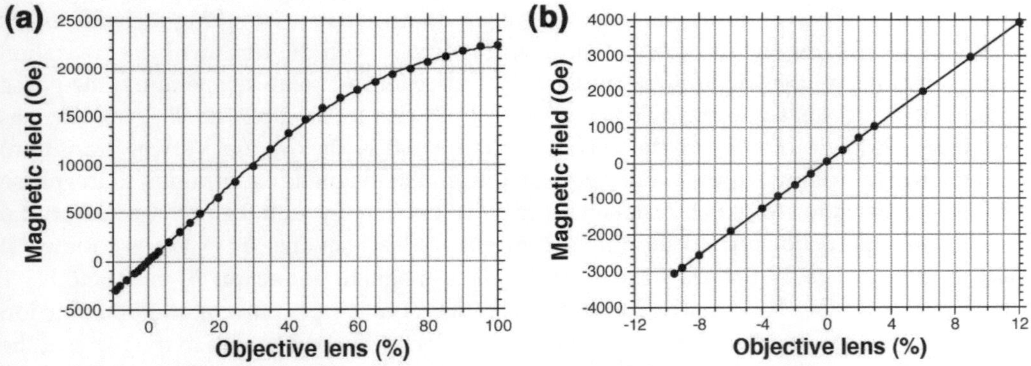

Fig. 6. Hall probe calibration of the magnetic field in the specimen plane for an FEI Titan 80-300ST TEM operated at 300 kV, plotted as a function of normalized objective lens current. In this condition, zero field is obtained at an objective lens current of -0.124%.

3. Applications of electron holography to magnetic materials

3.1 Magnetic properties of magnetite

Magnetite (Fe_3O_4) was the first magnetic material to be exploited by humankind and is one of the most important magnetic minerals in geological and biological systems (Dunlop & Özdemir, 1997). Recently, magnetite nanoparticles have been identified as promising candidates for medical applications such as targeted drug delivery (Pankhurst et al., 2009).

Here, we present two examples of the characterization of the magnetic properties of magnetite using electron holography.

3.1.1 Low-temperature magnetic properties

The Verwey transition in magnetite, which is a first-order crystallographic phase transition, has been studied extensively since its discovery (Verwey, 1939) because of its enormous impact on the magnetic and other physical properties of the material (Walz, 2001). The Verwey transition is associated with an order-of-magnitude increase in magnetocrystalline anisotropy and a change in magnetic easy axis from cubic <111> to monoclinic [001] below ~120 K. Although numerous studies have suggested that magnetic domain walls in magnetite can interact with the ferroelastic twin walls that form at low temperature (e.g., Smirnov & Tarduno, 2002), no direct evidence for such interactions has previously been presented. Here, for low-temperature electron holography experiments, a liquid-nitrogen-cooled TEM specimen holder (~90 K) was used.

3.1.1.1 Single-domain magnetite

Experimental results obtained using electron holography from an isolated 50-nm-diameter single crystal of magnetite from a bacterial cell are shown in Fig. 7. Figure 7a shows a high-resolution TEM image of the crystal, which is elongated slightly in the [111] direction. The three-dimensional morphology of the same crystal, determined using high-angle annular dark-field electron tomography, is shown in Fig. 7b. Magnetic induction maps were recorded using electron holography with the crystal in magnetic-field-free conditions, both at room temperature (Fig. 7c) and at 90 K (Fig. 7d). The magnetic contribution to the phase shift was determined by performing a series of *in situ* magnetization reversal experiments. Both induction maps show uniformly-magnetized single domain states, including a characteristic return flux resembling that of an isolated magnetic dipole, with the remanent magnetization direction making a large angle to the applied field direction. At room temperature, the phase contours in the crystal make an angle of ~20° to its [111] elongation direction (Fig. 7c), whereas they make an angle of ~15° to the [111] direction at 90 K (below the Verwey transition) (Fig. 7d). Figure 7e shows a line profile generated from the magnetic contribution to the phase shift that was used to create Fig. 7c, taken along a line passing through the center of the crystal in a direction perpendicular to the phase contours. A least-squares fit of the experimental phase profile to Eqs 19 and 20 yielded an in-plane magnetic induction $B_\perp$ of 0.6(±0.12) T (Harrison et al., 2007). This value is consistent with the room temperature saturation induction of magnetite, suggesting that the magnetization direction of the particle lies in the plane of the specimen, close to the [121] crystallographic direction. This direction corresponds to the longest diagonal dimension of the particle, suggesting that shape anisotropy dominates the magnetic state of the crystal at room temperature. In contrast, the 90 K phase profile yielded a value for $B_\perp$ of 0.46(±0.09) T (Harrison et al., 2007). This value is lower than the saturation induction of magnetite at 90 K, suggesting that, at remanence, the magnetization direction in the crystal is tilted out of the plane by ~40° to the horizontal. This direction corresponds approximately to $[100]_{cubic}$ or $[001]_{cubic}$, as expected from the prediction that the $[001]_{monoclinic}$ easy axis can lie along any one of the original $\langle 100 \rangle_{cubic}$ directions and that the effect of magnetocrystalline anisotropy on the magnetic state of the crystal is dominant in the monoclinic phase below the Verwey transition (Dunlop & Özdemir, 1997). Ferroelastic twin walls are not observed in crystals of this size.

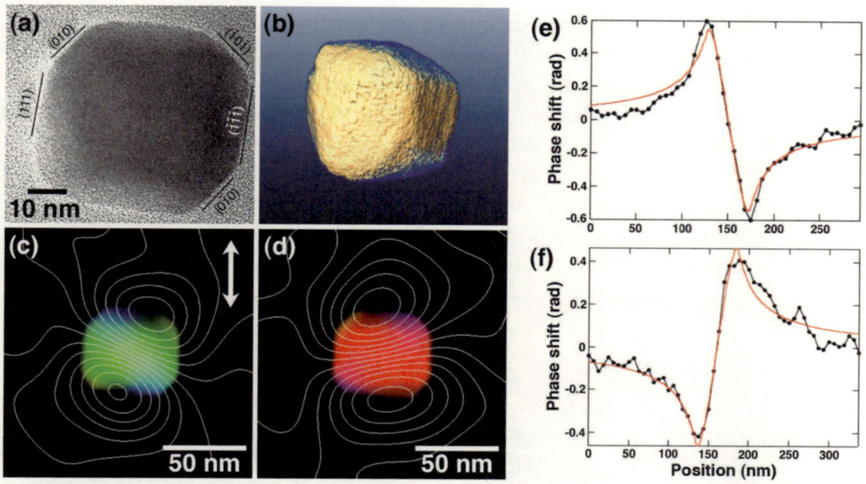

Fig. 7. (a) High-resolution TEM image of a 50-nm-diameter magnetite crystal from a magnetotactic bacterial cell (recorded by M. Pósfai). (b) Three-dimensional morphology of the same particle obtained using high-angle annular dark-field electron tomography (recorded by R.K.K. Chong). (c,d) Remanent magnetic states of the particle recorded at room temperature and at 90 K, respectively. The double arrow shows the applied field direction before reducing the experimental field to zero and recording holograms. (e,f) Data points in the line profiles show the magnetic contribution to the phase shift across the particle measured (e) at room temperature and (f) at 90 K. The red lines show least-squares fits to the data points, yielding values of $B_{\perp}$=0.60(± 0.12) T at room temperature and $B_{\perp}$=0.46(±0.09) T at 90 K.

3.1.1.2 Multi-domain magnetite

Results obtained from a synthetic magnetite specimen with a grain size of 10-30 μm, prepared for TEM using Ar ion milling, are shown in Fig. 8. Figure 8b shows a magnetic induction map recorded at room temperature in magnetic-field-free conditions using electron holography after applying a magnetic field of 2 T to the specimen. The presence of a magnetic flux-closure domain results from the minimization of magnetostatic energy. The magnetic state of this specimen below the Verwey transition (Fig. 8c) is significantly different from that observed at room temperature. At low temperature, the magnetic domains are more complicated and their sizes can be as small as several tens of nm. Magnetic domains that are far from the specimen edge have widths of 100-500 nm and are separated by 180° domain walls, which result from the strong uniaxial magnetocrystalline anisotropy of the monoclinic phase (antiparallel magnetization directions are shown in purple and yellow on the right side of Fig. 8c), as predicted in many previous studies (e.g., Dunlop & Özdemir, 1997).

The formation of magnetic domains that are related to "strain-contrast-free" ferroelastic twin walls, which are invisible in conventional TEM images, is depicted in Fig. 8d. The magnetic domain walls appear reproducibly at the same positions after applying large fields to the specimen in opposite directions, suggesting that the positions of the domain walls are defined strictly by underlying crystallographic features. Figure 8e shows a corresponding simulated magnetic induction map generated by using the approach of Beleggia & Zhu (2003), superimposed onto an experimental Lorentz (out-of-focus) TEM image. The crystallographic *c*-axis direction in each domain was determined using electron diffraction, revealing that neighboring monoclinic *c*-axes are at 90° to each other. Despite the fact that the parameters used to describe the specimen in the simulation were slightly different from its true geometry, the simulated image is in good agreement with the experimental magnetic induction map. A small remaining inconsistency in the direction of the phase contours in the purple domains may be caused by specimen thickness variations, stray magnetic fields and/or unknown additional phase ramps in the experimental image. Given that the $[001]_{monoclinic}$ easy axis direction in one of the twin domains points at ~45° out of the specimen plane, the presence of stray fields above and below the sample plane is to be expected. Magnetic structures such as those shown in Fig. 8 are affected significantly by the presence of ferroelastic twins, which therefore play a significant role in determining the unique magnetic properties of magnetite at low temperature (Kasama et al., 2010).

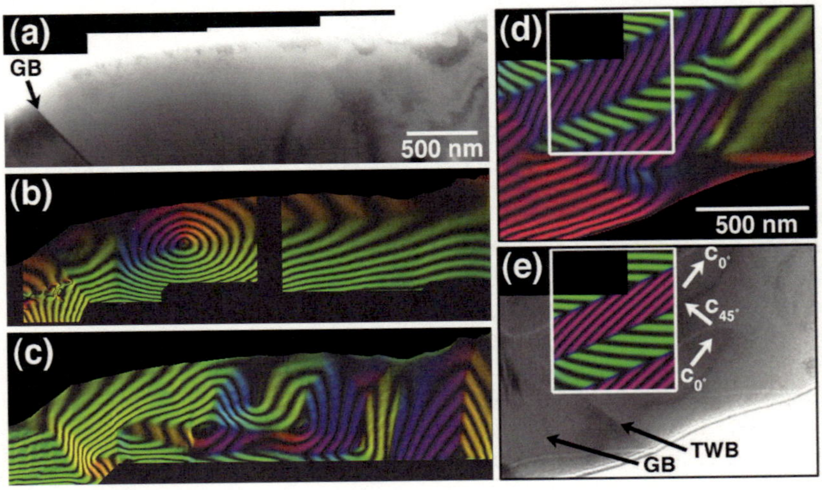

Fig. 8. (a) Bright-field TEM image of a synthetic magnetite specimen. (b,c) Remanent magnetic states recorded using off-axis electron holography at room temperature and below the Verwey transition, respectively. (d) Experimental magnetic induction map recorded from a region containing "strain-contrast-free" ferroelastic twin walls. (e) Simulated induction map superimposed onto an experimental Lorentz TEM image. The parameters used for the simulation were: B=0.6 T, V_0=17 V and a ferroelastic twin domain width of 165 nm. The specimen was assumed to be wedge-shaped with a thickness gradient of 10%. "$c_{0°}$" and "$c_{45°}$" refer to monoclinic c-axes that are oriented in-plane and at an angle of 45° to the specimen plane, respectively. The contour spacing is 2π. GB: grain boundary; TWB: twin wall boundary.

3.1.2 Critical size for superparamagnetic behavior in three-dimensional assemblies of magnetite nanoparticles

The particle or grain size is one of the most important parameters that determines the magnetic properties of a material. The effect of particle size and shape on the magnetic state of an isolated magnetite particle was discussed theoretically by Butler & Banerjee (1975), who suggested that the transition from single-domain to superparamagnetic behavior occurs below a particle size of 25-30 nm. Figure 9 shows local magnetic structures of magnetite particle assemblies on smectite (clay) platelets, which provide an ideal geometry for the experimental study of the critical threshold size for superparamagnetic behavior in the presence of strong interactions between adjacent particles (Galindo-Gonzalez et al., 2009). Electron holograms were acquired at remanence using a liquid-nitrogen-cooled TEM specimen holder. Unwanted mean inner potential contributions to the phase were eliminated by using *in situ* magnetization reversal. Figure 9a shows magnetite particles with an average diameter of ~10 nm, alongside two larger magnetite particles with diameters of 47 and 48 nm. The two larger particles are in close proximity to each other and result in a dipole-like magnetic signal (Fig. 9b). The measured induction in the particles is 0.53 T, which is close to that expected for magnetite (~0.6 T). The smaller particles do not show a detectable magnetic signal. Two agglomerates, each of which contains two touching particles of different size, are shown in Figs 9c and e. The particles in Fig. 9c have diameters of 33 and 42 nm, whereas in Fig. 9e they have diameters of 25 and 32 nm. The agglomerate containing the larger particles shows a magnetic signal (Fig. 9d), whereas the agglomerate with the smaller particles does not (Fig. 9f). This difference provides an estimate of 30–40 nm for the critical threshold size for superparamagnetic behavior in three-dimensional assemblages of magnetite over a timescale of ~10 sec (the time required for hologram collection). It should be noted that stray magnetic fields from larger particles can stabilize the moment of an adjacent smaller particle. An example of such behavior is shown in Figs 9g and h, where the stray field (the return flux) from larger particles appears to stabilize the moments of smaller particles, in the form of a flux-closed magnetic state that would show negligible overall magnetic remanence in a bulk measurement.

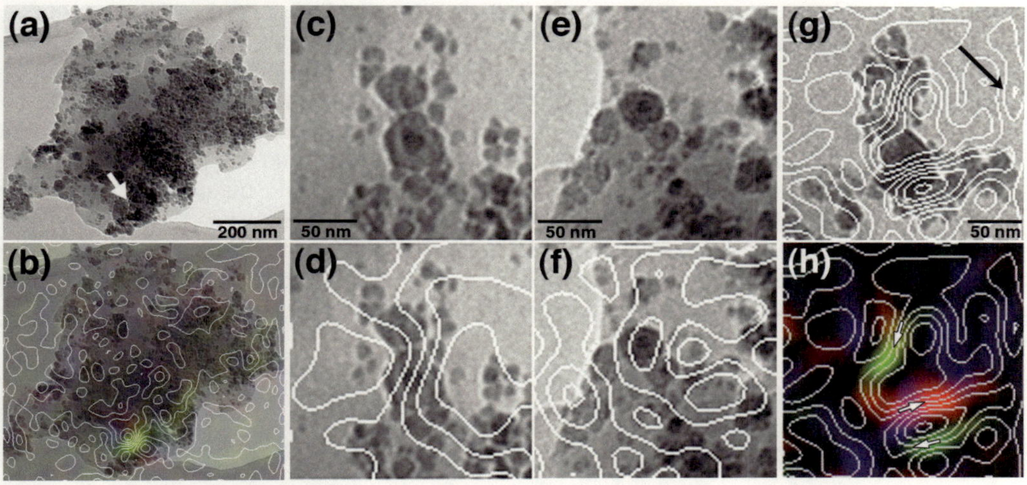

Fig. 9. (a) Bright-field TEM image of magnetite particles adhering to a smectite particle. (b) Corresponding magnetic induction map acquired at 92 K. Two larger crystals next to each other are marked by an arrow in (a). (c-f) Bright-field TEM images of assemblies containing larger, touching particles and corresponding remanent magnetic states. The larger particles in (c) and (d) have diameters of 33 and 42 nm, whereas those in (e) and (f) have diameters of 25 and 32 nm. (g) Magnetic phase contour and color maps recorded at remanence after applying a large field to the specimen in the direction shown with the black arrow in (g). The contour spacing in (b) is 0.125 radians, whereas those in (d), (f), (g) and (h) are 0.0625 radians. Adapted from Galindo-Gonzalez et al. (2009).

3.2 Magnetic thin films

Magnetic thin films are central to the development of spintronic devices such as magnetic random access memories and magnetoresistive read heads. For their development, it is important to characterize their magnetic states and switching field distributions, especially when they take the form of deep-submicron structures. However, most magnetometry techniques lack the sensitivity to characterize individual nanostructures. For example,

although magnetic force microscopy can be used to image stray magnetic fields above samples, it is invasive and cannot be used to measure the magnetization within them.
Here, we illustrate the application of electron holography to the characterization of the magnetic properties of thin films with nanometer spatial resolution. These studies have involved the design and application of a three-contact biasing TEM specimen holder, which allows magnetic devices to be examined under an electrical applied bias in the TEM (Kasama et al., 2005a).

3.2.1 Pseudo-spin-valve thin film elements

A magnetoresistive multilayer typically contains a magnetically hard pinned layer and a magnetically soft free layer, separated by a tunnel barrier or a conducting spacer layer. The resistance of the structure depends on the relative magnetization directions of the two layers. This effect can be used to sense small magnetic fields, which rotate the magnetization direction of the free layer to store a data bit. The sample used in this study comprised a two-dimensional array of 75 × 280 nm pseudo-spin-valve elements, which were prepared from a polycrystalline $Ni_{79}Fe_{21}$ (4.1 nm)/Cu (3 nm)/Co (3.5 nm)/Cu (4 nm) film that had been sputtered onto an oxidized silicon substrate (Kasama et al., 2005b). Figure 10 shows three representative magnetic induction maps recorded from three adjacent elements using electron holography in magnetic-field-free conditions. Figures 10a–c were acquired after saturating the elements upward and then applying external fields with in-plane components of (a) 0 Oe, (b) 424 Oe and (c) 1092 Oe downward. The contours were generated directly from the magnetic contribution to the measured holographic phase shift, after subtracting the mean inner potential contribution from each recorded phase image using *in situ* magnetization reversal. Where the phase contours in the elements are closely-spaced, the directions of the two magnetic layers are inferred to be parallel, whereas where the contours are widely-spaced the magnetic layers are inferred to be antiparallel. The magnetic configurations do not have end domains or vortices, and magnetic interactions between the neighboring elements are not significant.

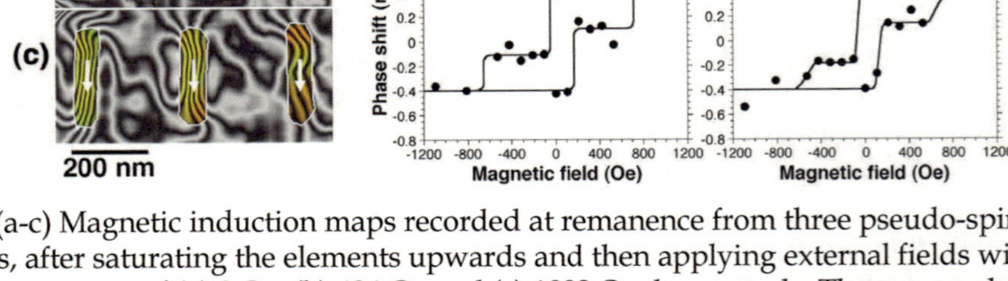

Fig. 10. (a-c) Magnetic induction maps recorded at remanence from three pseudo-spin-valve elements, after saturating the elements upwards and then applying external fields with in-plane components of (a) 0 Oe, (b) 424 Oe and (c) 1092 Oe downwards. The arrows show the local magnetization directions. The phase contour spacing is 0.098 radians. (d-g) Remanent hysteresis loops measured from each of the three elements shown in (a)-(c). The loops in (d), (e), (f) and (g) were obtained from the left, middle and right elements and their average, respectively. Adapted from Kasama et al. (2005b).

Remanent hysteresis loops measured from the magnetic contributions to the recorded phase images for each of the three elements are shown in Figs 10d-g. Figure 10d shows an antiparallel state in only one direction, with switching fields for the two layers of 100-200 and 100-400 Oe. In comparison, the other elements have switching fields of 100–200 and 500-700 Oe. Such switching variability is consistent with a collective hysteresis loop measured from ~10^9 elements, in which the switching field variations of the NiFe and Co layers are 50 and 150 Oe, respectively (Kasama et al., 2005b). The width of the switching field distribution is thought to result from variations in the shapes and sizes of the elements and from microstructural variability, which must be minimized if such elements are to be used in ultra-high-areal-density magnetic recording applications.

3.2.2 Magnetite thin film containing anti-phase domains

Magnetite is a promising candidate for spintronic applications because it is expected to be an efficient source of spin-polarised electrons. In layered magnetite films, it has been suggested that unexpectedly low magnetoresistance may result from the presence of antiferromagnetic coupling at anti-phase domain boundaries (APBs) (Fig. 11a), which would randomize the spins of conduction electrons. We have used electron holography to examine magnetic remanent states in an (001) magnetite thin film that was grown epitaxially on MgO and contains APBs (Kasama et al., 2006a). The magnetic induction maps that are shown in Figs 11b and d were acquired after saturating the sample magnetically downward and then applying fields with components in the plane of the sample of 192 and 513 Oe upward, respectively. The colors indicate the direction of the measured induction, according to the color wheel shown in Fig. 11c. In contrast to the examination of patterned magnetic nanostructures and nanocrystals, the mean inner potential contribution did not have to be subtracted from each recorded phase image because the film thickness was uniform across the field of view. The thin red and blue lines on each magnetic induction map correspond to the positions of the APBs that are visible in dark-field images obtained using 220 and 131 reflections, respectively. White lines correspond to the APBs that are visible using both reflections.

Although the remanent magnetic states are highly complicated, two types of magnetic domain contrast can be seen (Fig. 11b): (i) coarse magnetic domains (typically 100-350 nm in size) and (ii) finer magnetic domains (typically 10-90 nm in size). Adjacent coarse domains differ in their average magnetization direction by approximately 90° or 180°. Each coarse domain contains fluctuations in magnetization direction that occur on a scale similar to the size of the anti-phase domains. In Fig. 11d, antiferromagnetic coupling is visible as a 180° change in the local magnetic induction direction across several APBs, as indicated in the boxed region. The APBs that exhibit antiferromagnetic coupling are typically associated with a strong out-of-plane component of the magnetization in the film. Figure 11e shows the variation of the inclination of the magnetization from the plane of the film, inferred from the measured in-plane component of the magnetic induction and plotted as a function of distance from a single APB in Fig. 11d. The local moments at the position of the APB are inclined at an angle of almost 90° to the plane of the specimen, gradually rotating back to the specimen plane over a distance of ~17 nm. Our observations suggest that the magnetic properties of the magnetite thin film are related strongly to local crystallographic features at the positions of APBs. However, antiferromagnetic coupling is not present at each APB and the low value of magnetoresistance is unlikely to be explained by their presence alone (Kasama et al., 2006a).

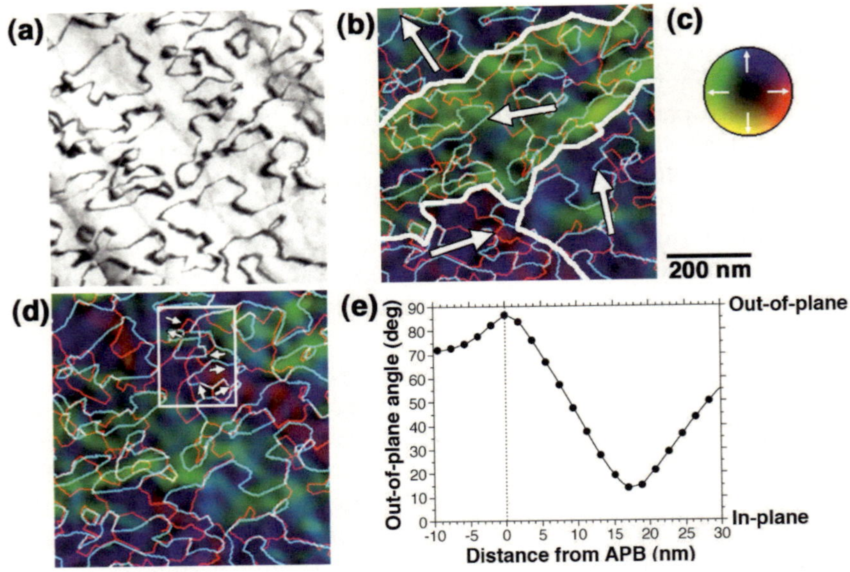

Fig. 11. (a) Dark-field image acquired using the 131 reflection of magnetite, showing APBs in a 25-nm-thick (001) magnetite thin film grown epitaxially on MgO. (b,d) Magnetic remanent states acquired from the region shown in (a), after saturating the film downward magnetically and then applying fields with in-plane components of 192 and 513 Oe upward, respectively. The arrows in (b) indicate the average magnetic induction direction in each coarse magnetic domain. The arrows in (d) show the local magnetic induction directions associated with antiferromagnetic coupling at APBs. The thin red, blue and white lines mark the positions of APBs. (c) Color wheel, representing the magnetic induction directions in (b) and (d). (e) Measured variation of the inclination of the magnetization from the plane of the film plotted as a function of distance from an APB. Adapted from Kasama et al. (2006a).

3.2.3 Current-induced motion of magnetic domain walls in permalloy wires

It has recently been demonstrated that spin-transfer effects associated with injected currents can be used to displace magnetic domain walls (Saitoh et al., 2004). This effect is of interest for the development of novel memory and logic devices based on domain-wall propagation (Allwood et al., 2002). A combination of Lorentz TEM and electron holography can be used to study the motion of magnetic domain walls induced by currents (Junginger et al., 2007) by using an electrical biasing TEM specimen holder, as described above. Here, permalloy zigzag line structures with a line width and thickness of 430 and 11 nm, respectively were fabricated on electron-transparent SiN windows using electron-beam lithography, as shown in Fig. 12a. Figure 12b shows the sequential positions of a domain wall that was subjected to 10 μs pulses with a current density of 3.14×10^{11} A/m^2. Electron holograms were acquired at each of these positions. It was observed that a transverse wall initially formed at a kinked region of the wire after the application of a magnetic field. After applying a current pulse, the domain wall moved by 2330 nm in the direction of electron flow and transformed into a vortex-type domain wall. After a second pulse, the vortex domain wall moved slightly in the same direction and became distorted (3 in Fig. 12b), with the long axis of the vortex inclined increasingly perpendicular to the wire length. This behavior may be associated with edge roughness or defects, which often restrict magnetic wall movements. After the third pulse, the domain wall moved by a further 260 nm and retained a vortex state. More recent studies have involved the injection of small currents to influence thermally-activated magnetic domain wall motion between closely-adjacent pinning sites (Eltschka et al., 2010).

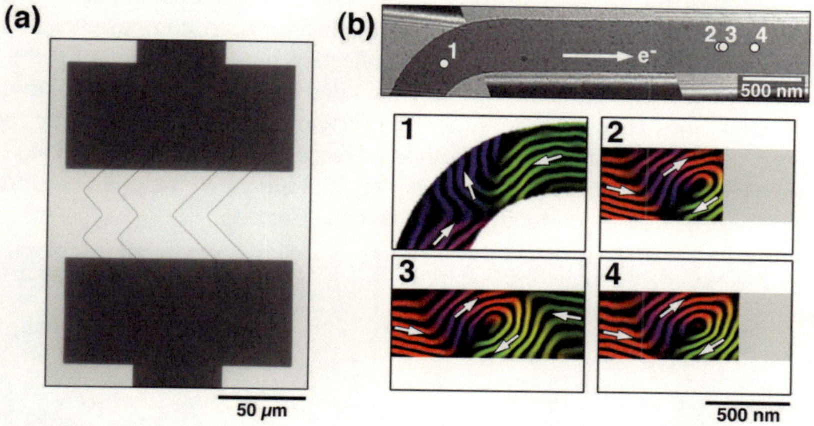

Fig. 12. (a) Low-magnification TEM image of four permalloy zigzag wires (width: 430 nm, thickness: 11 nm) patterned onto a SiN window using electron-beam lithography. A kink in one wire was examined using electron holography. (b) Higher-magnification TEM image of a permalloy wire, showing the motion of a domain wall after the injection of 10 μs pulses with a current density of 3.14×10^{11} A/m^2, showing magnetic induction maps obtained from a domain wall at each position. The long arrow in the top image indicates the direction of electron flow. The phase contour spacing is 0.785 radians.

3.3 Magnetotactic bacteria

Magnetotactic bacteria comprise a number of aquatic species that migrate along geomagnetic field lines. This behavior results from the presence of intracellular magnetic crystals of magnetite or greigite (Fe_3S_4). Similar magnetic crystals are also found in other organisms, such as fish, birds and humans, although their function is often poorly understood (Pósfai & Dunin-Borkowski, 2009). Here, we use electron holography to study the magnetic microstructures of greigite crystals in magnetotactic bacteria that contain multiple chains of either slightly elongated or almost equidimensional crystals (Kasama et al., 2006b).

A magnetic induction map recorded using off-axis electron holography from a dividing bacterial cell is shown in Figure 13. The cell contains several highly-elongated magnetite crystals alongside 50-nm-diameter greigite crystals. The measured magnetic induction approximately follows the distribution of greigite crystals in the multiple chain (Fig. 13). However, there are considerable variations in the local direction of the induction within individual particles, presumably as a result of variations in their morphology and orientation. In general, the more elongated crystals appear to be more strongly magnetic, particularly if the elongation and chain axes are parallel to each other. However, several crystals in the outer regions of the chain either appear to be non-magnetic or their induction elongation are parallel to the elongation of the crystal rather than to the chain axis. The crystals that appear to be non-magnetic may be magnetized largely out of the plane of the specimen, since electron holography is only sensitive to the in-plane components of the magnetic induction. Overall, shape anisotropy appears to be the most important factor controling the magnetization direction of each crystal (unless they are very closely-spaced), followed by interparticle interactions, whereas magnetocrystalline anisotropy is the least important. Although the crystals in the cell are less well organized with respect to their morphology, crystallographic orientation and position, when compared with magnetite-producing bacteria, they provide a measured collective magnetic moment (~9 x 10^{-16} Am^2) that is sufficient for magnetotaxis and is calculated to allow the cell to migrate along geomagnetic field lines at more than 90% of its forward speed. The magnetic moment and forward speed were determined using methods described by Dunin-Borkowski et al. (1998) and Frankel (1984), respectively. These results provide new insight for understanding magnetotaxis in iron sulfide-producing magnetotactic bacteria.

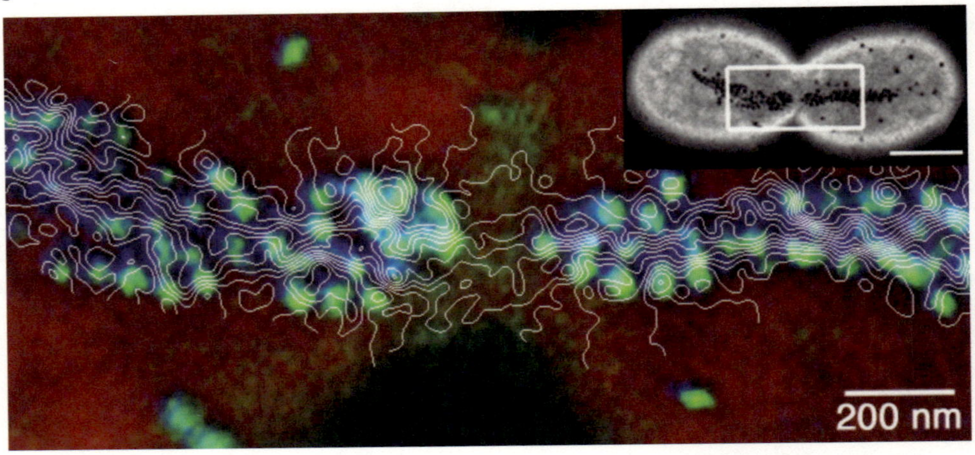

Fig. 13. Magnetic induction map recorded from greigite (Fe_3S_4) and magnetite (Fe_3O_4) crystals in a dividing magnetotactic bacterial cell using off-axis electron holography, superimposed onto an elemental map (red: carbon, green: sulfur, blue: iron). The rounded greigite crystals (light blue) are covered by an amorphous iron oxide (darker blue). The phase contour spacing is 0.098 radians. A carbon map of the entire dividing cell is shown in the inset (the scale bar is 1 μm). The box indicates the region from which the magnetic induction map was obtained. Adapted from Kasama et al. (2006b).

4. Quantitative measurement of magnetic moments using electron holography

The magnetic moment of a nanostructure is given by the expression

$$m = \iiint \mathbf{M}(\mathbf{r}) d^3 r \qquad (27)$$

where **M**(**r**) is the position-dependent magnetization and **r** is a three-dimensional position vector. Unfortunately, a holographic phase image does not provide direct information about **M**(**r**). Instead, the recorded phase shift is proportional to the projection (in the beam direction) of the in-plane components of the magnetic induction **B**(**r**) both within and around the specimen. We have recently shown mathematically that the magnetic moment of a nanostructure can in fact be measured quantitatively from a phase image or its gradient components (Beleggia et al., 2010). The basis of the algorithm is the relationship between the volume integral of the magnetic induction, a quantity that is referred to here as the "inductive moment" $\mathbf{m_B}$ and the magnetic moment. If a circular boundary around an isolated nanostructure of interest is chosen when integrating the phase shift in the form

$$\mathbf{m_B} = \frac{\hbar R}{e\mu_0} \int_0^{2\pi} d\theta [-\sin\theta, \cos\theta, 0] \phi(R\cos\theta, R\sin\theta) \tag{28}$$

where R is the radius of the integration circle, μ_0 is the vacuum permeability and θ is the polar angle. Then the following expression is found to be valid:

$$\mathbf{m} = 2\mathbf{m_B} \,. \tag{29}$$

Furthermore, two orthogonal components of $\mathbf{m_B}$ can be extrapolated as a function of the radius of the integration circle to a circle of zero radius to yield a measurement that is free of most artifacts (see Beleggia et al. (2010) for details).

In order to demonstrate the applicability of this approach to experimental phase images, we measured the magnetic moment of a chain of three closely-spaced magnetite crystals from a magnetotactic bacterium (Fig. 14a). The gradient of the magnetic contribution to the recorded phase shift is integrated within circles with radii of 100 to 240 nm in 1 nm increments (Fig. 14b). The two components of the inductive moment that are calculated from each measurement are extrapolated quadratically to zero circle radius, as shown in Fig. 14c. The best-fitting value for the measured inductive moment is 2.74(±0.18)×10^6 μ_B oriented at 136(±4)°. The magnetization obtained from the inductive moment and the volume of the three crystals is 510(±90) kA/m, which is in good agreement with that expected for magnetite, 480 kA/m.

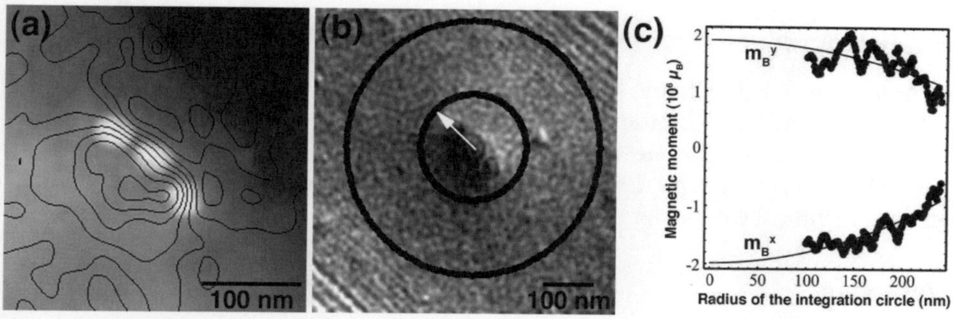

Fig. 14. (a) Magnetic induction map acquired using electron holography from three magnetite crystals in a bacterial cell. The phase contour spacing is 0.0625 radians. (b) Magnetic contribution to the phase shift corresponding to (a). The circles and arrow show integration radii of 100 and 240 nm and the direction of the measured inductive moment determined by extrapolating the measurements quadratically to zero integration radius, as shown in (c). Adapted from Beleggia et al. (2010).

5. Conclusions

In combination with TEM techniques that allow crystallographic and compositional information about materials to be obtained at atomic resolution, off-axis electron

holography provides a powerful tool for the study of nanometer-scale magnetic fields in materials and devices, especially if advanced TEM specimen holders are used to apply external stimuli for studying processes such as current-induced domain wall motion.
The phase sensitivity and spatial resolution of off-axis electron holography have been discussed by de Ruijter & Weiss (1993), who suggested that a practical phase sensitivity of $2\pi/100$ radians can be achieved by using digital recording and image processing. In magnetic samples, this sensitivity corresponds to a detectable signal from a ~6 nm particle with a magnetization similar to that of cobalt (1.8 T). In the future, a phase sensitivity of $2\pi/1000$ radians and a spatial resolution of a few nanometers may be required for studies of spintronic devices such as diluted magnetic semiconductors and self-assembled nanostructured devices. Such experiments will require the use of high-brightness electron sources, aberration correctors, highly sensitive recording media, drift-free specimen stages and holders and sophisticated software for the automation of lengthy experiments. In addition, novel image analysis and model-based algorithm will be necessary to enhance weak magnetic signals.
One of the major limitations of electron holography is that it is currently only able to image the *projection* of the in-plane magnetic field onto the two-dimensional sample plane. The combination of electron holography with electron tomography would allow three-dimensional magnetic vector fields inside nanostructured materials to be visualized by applying suitable reconstruction techniques to ultra-high-tilt series of electron holograms acquired about orthogonal specimen-tilt axes (Lai et al., 1994). The practical achievement of magnetic vector field tomography is an attractive prospect, as it would lead to the complete characterization of the magnetic field inside real nanostructured materials and devices that have three-dimensional shapes and distributions.

6. Acknowledgment

We are grateful to R.K.K. Chong, N.S. Church, J.M. Feinberg, R.J. Harrison, J.R. Jinschek, M. Kläui, E.T. Simpson, M. Pósfai and other colleagues for the provision of samples and for ongoing collaborations. The electron holography described here was carried out at the Electron Microscopy group, University of Cambridge and the Center for Electron Nanoscopy, Technical University of Denmark.

7. References

Allwood, D.A.; Xiong, G.; Cooke, M.D.; Faulkner, C.C.; Atkinson, D.; Vernier, N. & Cowburn, R.P. (2002). Submicrometer ferromagnetic NOT gate and shift register. Science 296, 2003-2006.

Beleggia, M. & Zhu, Y. (2003). Electron-optical phase shift of magnetic nanoparticles: I. Basic concepts. Philos. Mag. 83, 1045–1057.

Beleggia, M.; Kasama, T. & Dunin-Borkowski, R.E. (2010) The quantitative measurement of magnetic moments from phase images of nanoparticles and nanostructures - I. Fundamentals. Ultramicroscopy 110, 425-432.

Butler, R.F. & Banerjee, S.K. (1975). Theoretical single-domain grain size range in magnetite and titanomagnetite. J. Geophys. Res. 80, 4049-4058.

Cooper, D.; Truche, R.; Rivallin, P.; Hartmann, J-M.; Laugier, F.; Bertin, F.; Chabli, A. & Rouviere, J-L. (2007). Medium resolution off-axis electron holography with millivolt sensitivity. Appl. Phys. Lett. 91, 143501.

Cowley, J.M. (1992). Twenty forms of electron holography. Ultramicroscopy 41, 335-348.

Cowley, J.M. & Spence, J.C.H. (1998). Principles and theory of electron holography, In: Introduction to Electron Holography, E. Völkl, L.F. Allard & D.C. Joy (Eds), pp.17-56, Kluwer Academic/Plenun Publishers, New York.

de Graef, M., Nuhfer NT, McCartney, MR. (1999) Phase contrast of spherical magnetic particles. J. Microsc. 194, 84-94.

de Ruijter, W.J. & Weiss, J.K. (1993) Detection limits in quantitative off-axis electron holography. Ultramicrscopy 50, 269-283.

Dunin-Borkowski, R.E.; McCartney, M.R.; Frankel, R.B.; Bazylinski, D.A.; Pósfai, M. & Buseck, P.R. (1998). Magnetic microstructure of magnetotactic bacteria by electron holography. Science 282, 1868-1870.

Dunin-Borkowski, R.E.; McCartney, M.R. & Smith, D.J. (2004). Electron holography of nanostructured materials, In: H.S. Nalwa (Ed.), Encyclopedia of Nanoscience and Nanotechnology, Vol. 3, pp.41-100, American Scientific Publishers.

Dunlop, D.J. & Özdemir, Ö. (1997). Rock magnetism: Fundamentals and frontiers, Cambridge University Press, Cambridge.

Egerton, R.F. (1996). Electron energy-loss spectroscopy in the electron microscope (2nd edition), Plenum, New York.

Eltschka, M.; Wötzel, M.; Rhensius, J.; Krzyk, S.; Nowak, U.; Kläui, M.; Kasama, T.; Dunin-Borkowski, R.E.; Heyderman, L.J.; van Driel, H. J. & Duine, R.A. (2010). Nonadiabatic spin torque investigated using thermally activated magnetic domain wall dynamics. Phys. Rev. Lett. 105, 056601.

Erni, R.; Rossell, M.D.; Kisielowski, C. & Dahmen, U. (2009). Atomic-resolution imaging with a sub-50-pm electron probe. Phys. Rev. Lett., 102, 096101.

Frankel, R.B. (1984). Magnetic guidance of organisms. Ann. Rev. Biophys. Bioeng. 13, 85-103.

Frost, B.G.; Völkl, E & Allard, L.F. (1996). An improved mode of operation of a transmission electron microscope for wide field off-axis holography. Ultramicrscopy 63, 15-20.

Gabor, D. (1949). Microscopy by reconstructed wave-fronts. Proc. Roy. Soc. A 197, 454-487.

Galindo-Gonzalez, C.; Feinberg, J.M.; Kasama, T.; Cervera-Gontard, L.; Pósfai, M.; Kósa, I.; Duran, J.D.G.; Gil, J.E.; Harrison, R.J. & Dunin-Borkowski, R.E. (2009). Magnetic and microscopic characterization of magnetic nanoparticles adhered to clay surfaces. Am. Mineral. 94, 1120-1129.

Haine, M.E. & Mulvey, T. (1952). The formation of the diffraction image with electrons in the Gabor diffraction microscope. J. Opt. Soc. Am. 42, 763-769.

Harada, K.; Akashi, T.; Togawa, Y.; Matsuda, T. & Tonomura, A. (2005). Optical system for double-biprism electron holography. J Electron Microsc. 54, 19-27.

Harada, K.; Matsuda, T.; Tonomura, A.; Akashi, T. & Togawa, Y. (2006). Triple-biprism electron interferometry. J. Appl. Phys. 99, 113502.

Harrison, R.J.; Dunin-Borkowski, R.E.; Kasama, T.; Simpson, E.T. & Feinberg, J.M. (2007). Magnetic properties of rocks and minerals, In: Treatise on Geophysics, Vol. 2, G. Schubert (Ed.), pp.579-630, Elsevier Science.

Junginger, F.; Kläui, M.; Backes, D.; Rüdiger, U.; Kasama, T.; Dunin-Borkowski, R.E.; Heyderman, L.J.; Vaz, C.A.F. & Bland, J.A.C. (2007). Spin torque and heating effects in current-induced domain wall motion probed by transmission electron microscopy. Appl. Phys. Lett. 88, 212510.

Kasama, T.; Dunin-Borkowski, R.E.; Matsuya, L.; Broom, R.F.; Twitchett, C.A.; Midgley, P.A.; Newcomb, S.B.; Robins, A.C.; Smith, D.W.; Gronsky, J.J.; Thomas, C.A. & Fischione, P.E. (2005a). A versatile three-contact electrical biasing transmission electron microscope specimen holder for electron holography and electron tomography of working devices. In In-situ Electron Microscopy of Materials, Vol. 907E, P.J. Ferreira, I.M. Robertson, G. Dehm, and H. Saka (Eds), Materials Research Society, Pittsburgh, PA, pp. MM13-02.

Kasama, T.; Barpanda, P.; Dunin-Borkowski, R.E.; Newcomb, S.B.; McCartney, M.R.; Castaño, F.J. & Ross, C.A. (2005b). Off-axis electron holography of individual pseudo-spin-valve thin film magnetic elements. J. Appl. Phys. 98, 013903.

Kasama, T.; Dunin-Borkowski, R.E. & Eerenstein, W. (2006a). Off-axis electron holography observation of magnetic microstructure in a magnetite (001) thin film containing antiphase domains. Phys. Rev. B 73, 104432.

Kasama, T.; Pósfai, M.; Chong, R.K.K.; Finlayson, A.P.; Buseck, P.R.; Frankel, R.B. & Dunin-Borkowski, R.E. (2006b). Magnetic properties, microstructure, composition and morphology of greigite nanocrystals in magnetotactic bacteria from electron holography and tomography. Amer. Mineral. 91, 1216-1229.

Kasama, T.; Dunin-Borkowski, R.E.; Scheinfein, M.R.; Tripp, S.L.; Liu, J. & Wei, A. (2007). Off-axis electron holography of self-assembled Co nanoparticle rings. In: Quantitative Electron Microscopy for Materials Science, Vol. 1026E, E. Snoeck, R.E. Dunin-Borkowski, J. Verbeeck & U. Dahmen (Eds), Materials Research Society, Warrendale, PA, pp. C18-03.

Kasama, T.; Dunin-Borkowski, R.E.; Scheinfein, M.R.; Tripp, S.L.; Liu, J. & Wei, A. (2008). Reversal of flux closure states in cobalt nanoparticle rings with coaxial magnetic pulses. Adv. Mater. 20, 4248-4252.

Kasama, T.; Church, N.; Feinberg, J.M.; Dunin-Borkowski, R.E. & Harrison, R.J. (2010). Direct observation of ferromagnetic/ferroelastic domain interactions in magnetite below the Verwey transition. Earth Planet. Sci. Lett. 297, 10-17.

Lai, G.; Hirayama, T.; Ishizuka, K. & Tonomura, A. (1994). Three-dimensional reconstruction of electric-potential distribution in electron-holographic interferometry, J. Appl. Opt. 33, 829–833.

Lehmann, M.; Lichte, H.; Geiger, D.; Lang, G. & Schweda, E. (1999). Electron holography at atomic dimensions – present state. Mater. Character. 42, 249-263.

Lichte, H. (1991). Electron image plane off-axis electron holography of atomic structures. Adv. Opt. Electron Microsc. 12, 25-91.

Lichte, H. (1996). Electron holography: optimum position of the biprism in the electron microscope. Ultramicroscopy 64, 79-86.

Lichte, H. (2008). Performance limits of electron holography. Ultramicroscopy 108, 256-262.

Loudon, J.C.; Mathur, N.D. & Midgley, P.A. (2002). Charge-ordered ferromagnetic phase in $La_{0.5}Ca_{0.5}MnO_3$. Nature 420, 797-800.

Matsuda, T.; Hasegawa, S.; Igarashi, M.; Kobayashi, T.; Naito, M.; Kajiyama, H.; Endo, J.; Osakabe, N.; Tonomura, A. & Aoki, R. (1989). Magnetic field observation of a single flux quantum by electron-holographic interferometry. Phys. Rev. Lett. 62, 2519-2522.

Möllenstedt, G. & Düker, H. (1956). Beobachtungen und messungen an biprisma-interferenzen mit electronenwellen. Z. Physik 145, 377-397.

Pankhurst, Q.A.; Thanh, N.K.T.; Jones, S.K. & Dobson, J. (2009) Progress in applications of magnetic nanoparticles in biomedicine. J. Phys. D: Appl. Phys. 42, 224001.

Phatak, C.; Bain, J.A.; Zhu, J.G. & De Graef, M. (2008). Aberration corrected Lorentz microscopy for perpendicular magnetic recording media. Microsc. Microanal. 14, 832-833.

Pósfai, M & Dunin-Borkowski, R.E. (2009). Magnetic nanocrystals in organisms. Elements 5, 235-240.

Saitoh, E.; Miyajima, H.; Yamaoka, T. & Tatara, G. (2004). Current-induced resonance and mass determination of a single magnetic domain wall. Nature 432, 203-206.

Sickmann, J.; Formanek, P.; Linck, M.; Muehle, U. & Lichte, H. (2011). Imaging modes for potential mapping in semiconductor devices by electron holography with improved lateral resolution. Ultramicroscopy 111, 290-302.

Smirnov, A.V. & Tarduno, J.A. (2002). Magnetic field control of the low temperature magnetic properties of stoichiometric and cation-deficient magnetite. Earth Planet. Sci. Lett. 194, 359-368.

Smith, D.J. (2007). Characterisation of nanomaterials using transmission electron microscopy, In: Nanocharacterisation, A.I. Kirkland & J.L. Hutchison (Eds), pp.1-27, RSC Publishing, Cambridge.

Smith, D.J. & McCartney, M.R. (1998). Practical electron holography, In: Introduction to Electron Holography, E. Völkl, L.F. Allard & D.C. Joy (Eds), pp.87-106, Kluwer Academic/Plenun Publishers, New York.

Takeda, M.; Ina, H. & Kobayashi, S. (1982). Fourier-transform method of fringe-pattern analysis for computer-bases topography and interferometry. J. Opt. Soc. Am. 72, 156-160.

Tonomura, A. (1992). Electron-holographic interference microscopy. Adv. Phys. 41, 59-103.

Tonomura, A; Osakabe, N; Matsuda, T.; Kawasaki, T.; Endo, J.; Yano, S. & Yamada, H. (1986). Evidence for Aharonov-Bohm effect with magnetic field completely shielded from electron wave. Phys. Rev. Lett. 56, 792-795.

Verbeeck, J.; Bertoni, G. & Lichte, H. (2011). A holographic biprism as a perfect energy filter? Ultramicroscopy 111, 887-893.

Verwey, E.J.W. (1939). Electronic conduction of magnetite (Fe_3O_4) and its transition point at low temperature. Nature 144, 327-328.

Völkl, E & Lichte, H. (1990). Electron holograms for subångström point resolution. Ultramicroscopy 32, 177-180.

Völkl, E.; Allard, L.F.; Datye, A. & Frost, B.G. (1995). Advanced electron holography: a new algorithm for image processing and a standardized quality test for the FEG electron microscope. Ultramicroscopy 58, 97-103.

Walz, F. (2002). The Verwey transition - a topical review. J. Phys. Cond. Matter. 14, R285-R340.

Yamamoto, K.; Hirayama, T. & Tanji, T. (2004). Off-axis electron holography without Fresnel fringes. Ultramicroscopy 101, 265-269.

4

Computational Seismic Holography of Acoustic Waves in the Solar Interior

Charles Lindsey[1], Douglas Braun[1], Irene González Hernández[2] and Alina Donea[3]
[1]*NorthWest Research Associates*
[2]*National Solar Observatory*
[3]*Monash University*
[1,2]*USA*
[3]*Australia*

1. Introduction

1.1 Helioseismology

The advent of solar seismology is widely recognized as the independent discoveries by Leighton, Noyes & Simon (1963) and Evans, Michard & Servajean (1962) of Doppler oscillations in the Sun's surface, mostly with periods ranging from about 2 to 8 minutes. This was recognized as the surface signature of waves traveling beneath the Sun's surface. These waves, now understood to be generated by convection a few hundred km beneath the Sun's surface (Stein et al. , 2004), penetrate deep beneath the surface, filling the solar interior. The idea of using observations of these waves as a diagnostic of the Sun's interior structure was introduced by Ulrich (1970), and developed at length by Deubner (1975) and Rhodes, Ulrich & Simon (1977). Continued development followed many different avenues, some of these similar in some ways to geoseismic diagnostics of the Earth's interior. Indeed, helioseismic holography and "migration theory", the latter developed by Claerbout (1970) for applications in geoseismology, share basic concepts in wave optics in very similar contexts. However, solar seismology has overwhelming advantages, both in the quality, extent and uniformity of the observations and in the optical quality of the solar interior as an acoustic medium. Solar seismology gave us maps of the solar interior rotation rate (Rhodes, Deubner & Ulrich , 1979), affirming that the Sun is a differentially rotating fluid, its convection zone continuously being warped as the equatorial region rotates significantly faster than the inner polar region and the outer equatorial convection zone rotates faster than the deep convection zone. Helioseismology is also the significant observational basis of our present understanding of the thermal structure of the solar interior in standard models such as Christensen-Dalsgaard, Proffitt & Thompson (1993).

Solar seismology has been largely developed along two significantly separate lines. What is now generally recognized as "global helioseismology" views the Sun as a system of harmonic oscillators and relies heavily on the frequencies of its thousands of normal modes to build a model of the general thermal structure of the Sun and how different layers of it rotate. These diagnostics give us significant discrimination in depth and some in latitude, but none in longitude. What is recognized as "local helioseismology" uses perspectives that tend to be more familiar to optics to focus on relatively compact regions. The local discrimination our

eyesight gives us is a benefit of a *phase coherence* that is preserved in the medium through which the light we see propagates. Coherence is a crucial benefit to all aspects of solar seismology as well.
Following developments by Zernike (1938) and van Cittert (1939), coherence has come to be formally expressed by a complex "mutual coherence function",

$$\Gamma(\mathbf{r},\, \mathbf{r}',\, \tau) \;=\; \langle \psi(\mathbf{r},\, t)\psi^{*}(\mathbf{r}',\, t\, +\, \tau)\rangle, \tag{1}$$

correlating complex wave amplitudes $\psi(\mathbf{r},\, t)$ and $\psi(\mathbf{r}',\, t\, +\, \tau)$ at points $\mathbf{r}$ and $\mathbf{r}'$ in space, where the angular brackets indicate an averaging of their contents over time, t (see Born & Wolf , 1975a).[1] This has a strong analogy in a broad spectrum of diagnostic techniques in local helioseismology called "time-distance helioseismology" (Duvall et al. , 1993). Computational seismic holography is among these techniques, and can be expressed in terms of mutual coherence functions. The mutual coherence function, $\Gamma(\mathbf{r},\, \mathbf{r}',\, \tau)$, offers a statistical facsimile of ψ itself where optical technology is insufficiently fast to capture and record the temporal variations of ψ directly. The heart of this facsimile is in that $\Gamma(\mathbf{r},\, \mathbf{r}',\, \tau)$ obeys the same wave equation as ψ, in both $(\mathbf{r},\, \tau)$-space and $(\mathbf{r}',\, \tau)$-space.
A powerful technique in electromagnetic optics for deriving information on $\Gamma(\mathbf{r},\, \mathbf{r}',\, \tau)$—and fundamental to the development of electromagnetic holography—has been the superposition of electromagnetic fields, $\psi(\mathbf{r},\, t)$ and $\psi(\mathbf{r}',\, t')$, nominally destined to arrive at separates points, $\mathbf{r}$ and $\mathbf{r}'$, at separate times, t and t', onto the same region at the same time, so that they interfere. The resulting interference fringes, give us phase information about Γ that registers on a photographic plate, a medium normally sensitive only to intensity, not phase. The development of electromagnetic holography in the 1960s used this interference-based technique very effectively, benefiting from the long coherence lengths of monochromatic radiation from lasers.
The major practical difference between helioseismology and the electromagnetic optics familiar to electromagnetic holography is our ability, in the case of the former, to map the wave-mechanical field, ψ, in temporal detail as well as spatial, sampling it several times over its acoustic period, $2\pi/\omega$, over most of the Sun's visible hemisphere for weeks, even months, with infrequent interruptions. Also relevant is that acoustic radiation in the solar interior is highly polychromatic; its coherence length is less than the mean wavelength of the acoustic spectrum. However, having a clear temporal record of ψ covering a large fraction of the Sun's surface circumvents the need for a statistical facsimile of ψ, such as Γ, to determine phase-coherent extrapolations of either ψ or Γ into a wave-mechanical medium. The extrapolation can simply be applied to ψ directly, according to the laws of wave mechanics as we understand them. Moreover, we can do this not just at a single monochromatic frequency but over the entire acoustic spectrum accessible to helioseismic observations. This turns out to be a great benefit, since the times required to accumulate statistics at minutes-long helioseismic periods are much greater than for the femtosecond periods of visible light.

---

[1] In standard optics, the complex field amplitude $\psi(\mathbf{r},\, t)$ is an analytic extension of respective real amplitudes $\psi^{r}(\mathbf{r},\, t)$ that in principle might could be measured and recorded in terms of real values, the former obtained by inverting the Fourier transform, $\hat{\psi}(\mathbf{r},\, \omega)$, of $\psi^{r}(\mathbf{r},\, t)$, but truncating the negative frequencies, ω, hence,

$$\psi(\mathbf{r},\, t) \;\equiv\; \frac{2}{\sqrt{2\pi}} \int_{0}^{\infty} \hat{\psi}(\mathbf{r},\, \omega)\, e^{i\omega t}\, d\omega. \tag{2}$$

The analytic extension, then, is a complex extension of the real-valued amplitude of which the real-valued amplitude is the real part: $\psi^{r}(\mathbf{r},\, t) = \mathrm{Re}\{\psi(\mathbf{r},\, t)\}$.

Computational holography of helioseismic observations, then, is not based on any facsimile of the superposed monochromatic reference beams familiar to electromagnetic holography. The lack of such a facsimile has made it somewhat unfamiliar to some interpreters of electromagnetic holography. However, the object is the same: the extrapolation of some phase-coherent attribute of the wave-mechanical field, ψ, from the surface that samples it a macroscopic distance therefrom, based on a phase-informative record of the signature ψ manifests at the sampling surface. This idea was first proposed by Roddier (1975). It was later re-introduced by Lindsey & Braun (1990) as a proposed means of viewing solar activity in the Sun's far hemisphere, i.e., viewing it acoustically from the near hemisphere through the intervening interior medium. Further developments are described by Braun et al. (1992), Lindsey et al. (1996) and Lindsey & Braun (1997), the latter expressing seismic holography in the context of the more-recently introduced time-distance helioseismology (Duvall et al. , 1993; 1996), and by Chang et al. (1997), Braun & Lindsey (1998), Lindsey & Braun (2000a), Chou (2000), and Lindsey & Braun (2000b).
The technical ability to accomplish holographic extrapolations without recourse to the coherence function, Γ, does not detract in any way from the usefulness of Γ in solar seismology. On the contrary, the temporal discrimination helioseismic recordings of ψ give us allows us to compute $\Gamma(\mathbf{r}, \mathbf{r}', \tau)$ directly from ψ without recourse to superposition and interference on a medium that is sensitive only to intensity. This is a major element of time-distance helioseismology, upon which the development of helioseismic holography has drawn heavily (Lindsey & Braun , 1997; 2000b), and of which of helioseismic holography can now conveniently be regarded as substantially a chapter.

1.2 Basic principles of solar acoustic holography

The computational technique upon which helioseismic holography is based is best expressed in the context of a simple example, introduced by Lindsey & Braun (2000b). Consider an ideal acoustic medium, such as we suppose the solar interior to approach, with one or more monopolar acoustic emitters submerged into it. Figure 1 illustrates such a scenario with two such sources, monochromatic in this instance, at significantly different depths, representing surfaces of constant phase at a particular instant as wave fronts. The only visible manifestation of this acoustic field, $\psi(\mathbf{r}, t)$, is the disturbances that appear at the solar surface, $\mathcal{S}_0$, which generally first appear at the point on $\mathcal{S}_0$ directly above the source whose issue they represent and propagate outward along $\mathcal{S}_0$ therefrom. The motion of these disturbances across $\mathcal{S}_0$ is the basic content of helioseismic observations of the solar surface in such a scenario.
Suppose now that we have a detailed record of the foregoing disturbances over some domain, $\mathcal{R}$, of $\mathcal{S}_0$ for a period of at least a few oscillatory periods. Elementary computational acoustic holography consists of the following exercise:

1. applying these disturbances in some domain, $\mathcal{P} \in \mathcal{S}_0$, *in time reverse* to the surface, $\mathcal{S}_0$, of an *acoustic model* of the solar medium that itself is devoid of sources, absorbers or other significant anomalies,
2. allowing the model to propagate the resulting disturbances backwards into its interior, and
3. sampling the regressed acoustic field in some domain within the model.

Figure 2 illustrates steps 2 and 3 for the instance of a sampling domain that is a surface, $\mathcal{S}_z$, of constant depth, z, beneath $\mathcal{S}_0$. We call the time-reversed acoustic field, $H_+(\mathbf{r}, t)$, in the model the "coherent acoustic egression", in that $H_+(\mathbf{r}, t)$ represents a disturbance manifested by a wave, ψ, that had arrived at the solar surface, $\mathcal{S}_0$, with every apparent intention of egressing through it from the interior of the medium—such as in the case a medium that would provide

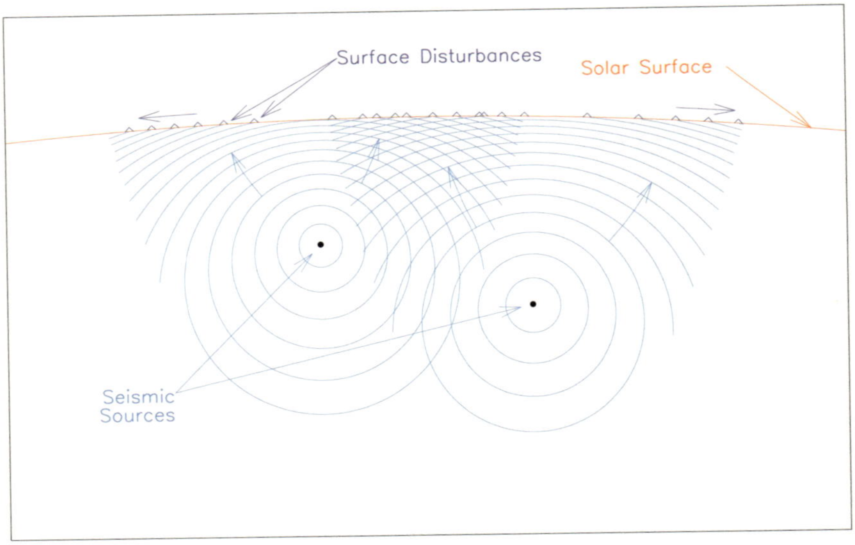

Fig. 1. Seismic waves emanating from submerged sources produce surface disturbances that propagate circularly outward along the solar surface from points directly above said sources, as indicated by generally out-going arrows.

somewhere for it to go after passing through $\mathcal{S}_0$, or in the case a surface boundary that would simply absorb it.

Once the submerged acoustic field, H_+, is recorded, a broad variety of applications are possible. The ideal analogy to what our eyes appear to give us, or a photographic plate, is the mean acoustic power. When such an averaging is applied directly to the acoustic field, ψ, we see at the solar surface, i.e., $\langle|\psi(\mathbf{r},\ t)|^2\rangle$, we call it an "acoustic-power map". Similarly applied to the acoustic egression, $\langle|H_+(\mathbf{r},\ t)|^2\rangle$, the result is an "egression-power map".

Based on our experience of optics, when the sampling surface sits at the depth of a source, $\langle|H_+(\mathbf{r},\ t)|^2\rangle$ should be characterized by a relatively compact signature, one that we recognize to be "in focus". This is supposedly the case for the source on the left in Figure 2, as represented by the egression-power plot at the bottom of the panel. If the sampling surface is moved substantially above or below the source depth, the signature remains extant, but spreads out of focus. Hence, when the sampling domain is a surface region, such as $\mathcal{S}_z$, we often call it "the focal surface". If $\mathcal{S}_z$ is planar in some approximation, we often call it the "focal plane". A particular point $\mathbf{r}$ in $\mathcal{S}_z$ can be called a "focal point", or simply a "focus" of the regression computation. The domain, $\mathcal{P}$, on $\mathcal{S}_0$ over which the observations, ψ, were applied in time reverse is called the "pupil" of the regression computation. In practice, it turns out to be fairly straight-forward to specify a pupil that is dependent upon the focus. This has strong advantages we will explain in the following section.

1.3 The computational task

A fundamental character of wave mechanics in a medium that preserves coherence is the possibility, at least in principle, of reconstructing, from measurements of a wave-mechanical field, ψ, in a thin surface, $\mathcal{S}$, significant information about ψ a long distance from $\mathcal{S}$—and an appropriately long time before or after the measurements. In relatively simple applications, this extrapolation is made in a medium that is relatively uniform, such as air. For simplicity, we will begin with a brief review of the concept for this case, and then extend the formalism

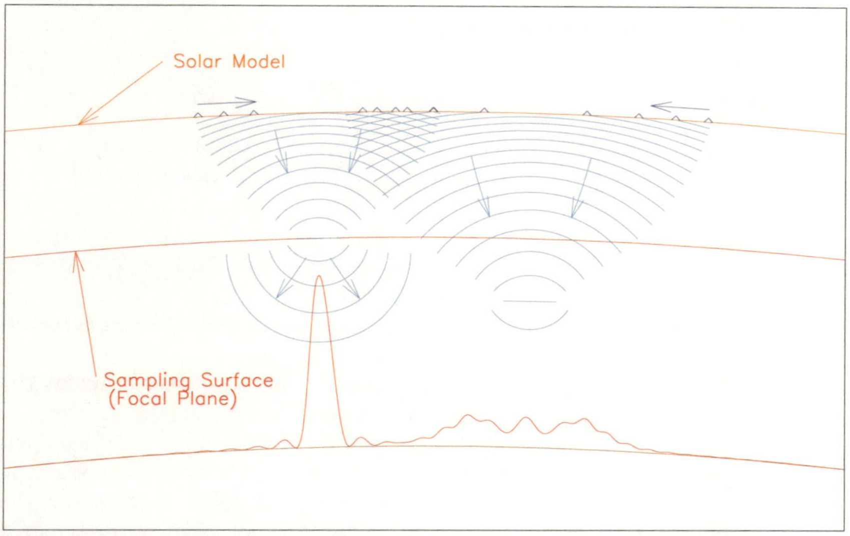

Fig. 2. Coherent computational regression of the surface acoustic field into the supposed solar interior. Surface disturbances recorded in the neighborhood overlying submerged sources are applied in time reverse to a sourceless acoustic model of the solar interior and computationally conducted back into the model interior. The underlying acoustic field differs in important respects from that actually produced by the sources. Nevertheless, a well-appropriated sampling of the regressed acoustic field renders localized sources with strong, compact signatures at appropriate depths. The seismic signature of a source that lies considerably below or above the sampling surface is rendered by a signature that is substantial but significantly out of focus.

for this to accommodate the significantly non-uniform solar interior. The former is the object of the Helmholtz-Kirchhoff integral theorem. For a formal elaboration of Helmholtz-Kirchhoff theory, we refer to Born & Wolf (1975b).

Provided an appropriately uniform medium described by a scalar wave-mechanical field, ψ, the Helmholtz-Kirchhoff integral prescribes $\psi(\mathbf{r},\, t)$ at any location, $\mathbf{r}$, in the interior of a closed region, V, and at any time, t, in terms of the values of ψ and its normal derivative, $\partial\psi/\partial n$, on the boundary, ∂V, of V in a range of "retarded times", $t_{\rm ret}$, appropriately prior to t:

$$\psi(\mathbf{r},\, t) \;=\; \int_{\partial V} d^2\mathbf{r}' \left(\frac{\partial}{\partial n'} \mathcal{G}_+(\mathbf{r},\, \mathbf{r}')\, \psi(\mathbf{r}',\, t_{\rm ret}) \;+\; \mathcal{G}_+(\mathbf{r},\, \mathbf{r}') \frac{\partial}{\partial n'}\, \psi(\mathbf{r}',\, t_{\rm ret}) \right), \tag{3}$$

where

$$\mathcal{G}_+(\mathbf{r},\, \mathbf{r}') \;=\; \frac{1}{4\pi|\mathbf{r} \,-\, \mathbf{r}'|}, \tag{4}$$

and

$$t_{\rm ret} \equiv t \;-\; \frac{|\mathbf{r} \,-\, \mathbf{r}'|}{c}, \tag{5}$$

with c representing the characteristic speed of propagation, supposed a constant. In preparation for the extension of this concept into a non-uniform medium, it is useful first to write equations (3–5) in the alternative form:

$$\psi(\mathbf{r},\, t) \;=\; \int_{-\infty}^{\infty} dt' \int_{\partial V} d^2\mathbf{r}' \left(\frac{\partial}{\partial n'}\, \mathcal{G}_+(\mathbf{r},\, \mathbf{r}';\, t \,-\, t')\;\psi(\mathbf{r}',\, t') \right.$$
$$\left. +\; \mathcal{G}_+(\mathbf{r},\, \mathbf{r}';\, t \,-\, t')\,\frac{\partial}{\partial n'}\,\psi(\mathbf{r}',\, t') \right), \tag{6}$$

where now

$$\mathcal{G}_+(\mathbf{r},\, \mathbf{r}';\, \tau) \;=\; \frac{1}{4\pi|\mathbf{r} \,-\, \mathbf{r}'|}\,\delta\Big(\tau \,-\, \frac{|\mathbf{r} \,-\, \mathbf{r}'|}{c}\Big), \tag{7}$$

in which δ signifies the Dirac-delta function, and $\mathcal{G}_+$ the "monopolar Green's function" of the medium.

A great deal of the practical substantiality of optics can be regarded in terms of a simple adaptation of the Helmholtz-Kirchhoff integral when nothing like the entirety of ψ *or* its normal derivative are accessible over anything like the entirety of any surface enclosing the specimen we propose to look at. For example, when we use our eyes to look at a luna moth ten meters from the surfaces of our corneas, the combined solid angle subtended by the two pupils of our eyes is only a few millionths of the 4π steradians over which the moth scatters radiation that hits it. It is well known that what our eyes give us is far from a complete representation of the electromagnetic field in the neighborhood of the moth, even in the relatively uniform electromagnetic medium exterior to the moth. We nevertheless learn a remarkable amount about moths from this incomplete representation. Based on this, we generally treat the incomplete representation such as that rendered by our limited pupils to be an important component of some aspect of the electromagnetic field in the neighborhood of the moth.

Helioseismic observations give us a representation of the acoustic field, ψ, in terms of the line-of-sight component of the motion of the medium, measured by the Doppler shift in a photospheric line over the Sun's near hemisphere. Hence, we propose to express the incomplete phase-coherent regression of ψ by the application of just the left term in the large parentheses in equation (6) over a limited region, $\mathcal{P}$, of the solar surface:

$$H_+(\mathbf{r},\, t) \;\equiv\; \int_{-\infty}^{\infty} dt' \int_{\mathcal{P}} d^2\mathbf{r}'\; G_+(\mathbf{r},\, \mathbf{r}';\, t \,-\, t')\,\psi(\mathbf{r}',\, t'), \tag{8}$$

where

$$G_+(\mathbf{r},\, \mathbf{r}';\, \tau) \;\equiv\; \frac{\partial}{\partial n'}\mathcal{G}_+(\mathbf{r},\, \mathbf{r}';\, \tau). \tag{9}$$

The point $\mathbf{r}$ is the aforementioned "focus" of the acoustic regression, introduced at the end of §1.2. The region, $\mathcal{P}$, over which the integral over $d^2\mathbf{r}'$ is taken is the "pupil" introduced directly thereafter. As mentioned at the end of §1.2, the pupil can be dependent upon the focus, $\mathbf{r}$. Indeed, it is generally useful to have the pupil "follow" the focus, keeping the same relative spatial relationship to it everywhere possible, so that effects such as diffraction will be as uniform as possible.

The extension of the Helmholtz-Kirchhoff formalism to a non-uniform acoustic medium entails two significant adaptations to uniform acoustics:

1. The non-uniformity of the medium is expressed by an appropriate revision of the Green's functions, $\mathcal{G}_+$ and G_+. In the case of a uniform medium, $\mathcal{G}_+$ could be—indeed was—expressed as a function of a single scalar, $|\mathbf{r} \,-\, \mathbf{r}'|$, greatly simplifying computational logistics. This is no longer so in solar-interior acoustics. In a spherically-symmetric

medium, however, both $\mathcal{G}_+$ and G_+ can be expressed as functions of the depth of **r** and the angle between **r** and surface **r**′ as projected from Sun center.

2. The solar interior is dispersive. Hence, the temporal dependencies of neither $\mathcal{G}_+$ nor G_+ are the infinitely sharp Dirac delta function expressed by equation (7).

For a discussion of means by which realistic Green's functions, $\mathcal{G}_+$ and G_+, can be determined in a medium such as the solar interior, we refer to Lindsey & Braun (2000b).
The computation of H_+ is greatly facilitated in the frequency domain of the temporal Fourier-transforms of the contestants, by the convolution theorem, which eliminates the integral over time. Thus,

$$\hat{H}_+(\mathbf{r}, \omega) = \int_{\mathcal{P}} d^2\mathbf{r}' \, \hat{G}_+(\mathbf{r}, \mathbf{r}'; \omega)\hat{\psi}(\mathbf{r}', \omega), \tag{10}$$

where $\hat{H}_+$, $\hat{G}_+$ and $\hat{\psi}$ signify the temporal Fourier transforms of H_+, G_+ and ψ, respectively. As mentioned in the context of Figure 2, it is often useful to render $H_+(\mathbf{r}, t)$ over a surface, $\mathcal{S}_z$, at a fixed depth, z, beneath $\mathcal{S}_0$. For that purpose, we express the location, **r**, of the focus by a single depth, z, and the point $\boldsymbol{\rho}$ overlying it on $\mathcal{S}_0$, which we equate to the unit sphere:

$$\mathbf{r} \equiv (\boldsymbol{\rho}, z). \tag{11}$$

This way, we can regard the derivation of H_+ on $\mathcal{S}_z$ from ψ on $\mathcal{S}_0$, under the specification of appropriate time intervals, regional domains and an appropriate pupil, to be the action of an operator, $P_+(z)$, applied to ψ:

$$H_+(\boldsymbol{\rho}, z, t) = P_+(z)\psi(\boldsymbol{\rho}, 0, t). \tag{12}$$

We call $P_+(z)$ the "(coherent acoustic) regression" operator under the foregoing specifications.

1.4 Subjacent-vantage holography

The diagrams shown in Figures 1 and 2 render the acoustic source as viewed acoustically from above it. This is called "superjacent-vantage" seismic holography. A major complication in solar-interior holography confronted in §1.3 is the acoustic non-uniformity of the solar model, even one that is devoid of local anomalies. The Sun's center is more than 2000 times the temperature of its surface. This is incumbent to a temperature that increases inexorably with depth, z. The sound speed, c, increases with depth accordingly, manifesting strong refraction. Optics in the ray approximation in the solar interior prescribe ray paths, according to Snell's law, whose incidences from vertical inexorably increase with depth. Hence, the solar interior landscape, if there were one, would appear very warped to an acoustic eye that was accustomed to optics in a uniform medium. Indeed, most of the acoustic radiation generated just beneath the surface is refracted back to the surface within a few tens of thousands of km of its source. This adds to the general complication introduced by the optical warpage in an important respect: The major acoustic sources are in essentially the same surface as the pupil—or only barely beneath it. The solar interior more than a few hundred km beneath the surface plays a role in helioseismology only for that component of the acoustic disturbance that propagates downward, penetrating to a significant depth beneath the surface before refracting back to the surface.
An important diagnostic option for this component of acoustic radiation, then, is to focus the computation at the very surface on which it is detected. Holography from this perspective is called "subjacent-vantage" holography. Figure 3 illustrates this application of solar seismic holography in an annular pupil surrounding the surface focus. In the familiar case of

a uniform medium, the ray paths would be straight, and prospective "acoustography" accomplished by a submerged acoustic camera as represented by the lens at the bottom of the Figure. Hence, subjacent-vantage holography of the surface shows the surface in the neighborhood of the focus as it would be seen looking up into it from beneath the surface.
For a rough electromagnetic analogy to subjacent-vantage holography, imagine a photographer using a mirror to photograph an insect on the lens of his camera. This may strike the reader as strange approach to entomology, but, the ability to focus our diagnostic on something we can directly see in electromagnetic radiation offers an exceptionally opportune control resource not to be left begging.

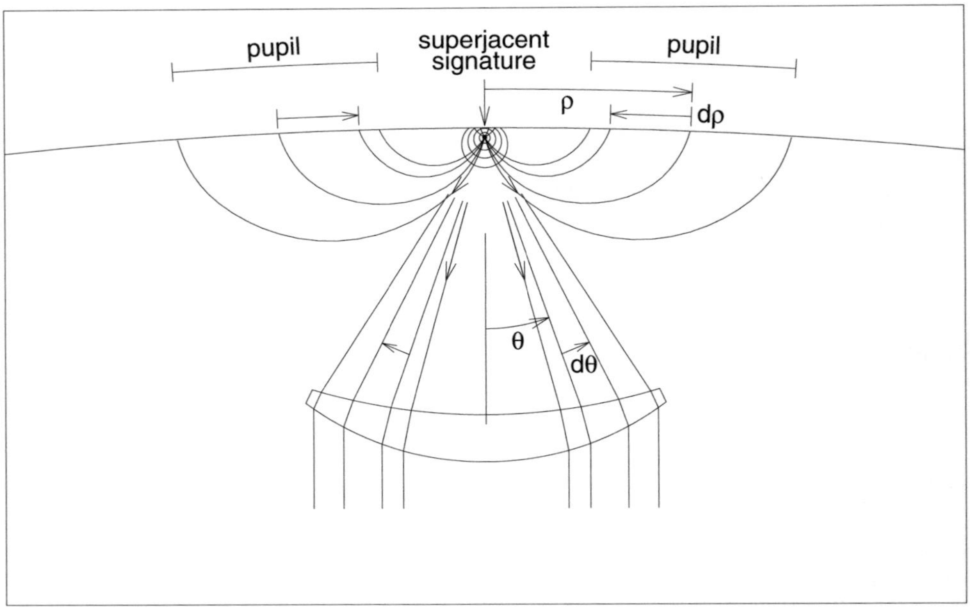

Fig. 3. Subjacent-vantage imaging is the result of a holographic regression in which the focal plane is shallow compared to the inner radius of the pupil. This configuration images seismic radiation that is initially emitted downward from the source and penetrates thousands of km into the solar interior before being refracted back to the surface. While the acoustic disturbance is necessarily observed at the surface, these images render the perspective of an acoustic observer looking upward into the base of the source from thousands of km beneath it. In subjacent-vantage holography, the computational pupil is substantially an inversion of that in familiar lens optics. As the angle, θ, of illumination at the focal point increases, the angular distance, ρ, along the pupil from its center, above the focal point, decreases rather than increases as it does in familiar lens optics.

Comparing the straight ray paths in Figure 3 that apply in a uniform medium with their curved counterparts, it will become apparent that the annular pupil applied in subjacent-vantage holography is somewhat of an inversion of that which applies in lens optics in a uniform medium. The ray directed toward the outer radius of the supposed lens comes to the surface at the inner radius of the annular pupil. Indeed, the diffraction limit of the reconstruction is primarily characteristic of the inner radius, the smaller that being the finer the resolution attained.

Figure 4 shows the result of subjacent-vantage egression-power holography applied to surface observations of a computational simulation of random acoustic disturbances generated just beneath the surface of an acoustic medium that conforms to the solar-interior model of Christensen-Dalsgaard, Proffitt & Thompson (1993). In this simulation, alpha-numeric absorbers, are placed (1) at the surface of the model, and (2) 56 Mm beneath the surface. It will be evident that absorbers play a role in some respects quite the opposite of emitters, casting acoustic silhouettes into the focal planes at their respective depths as opposed to positive signatures.

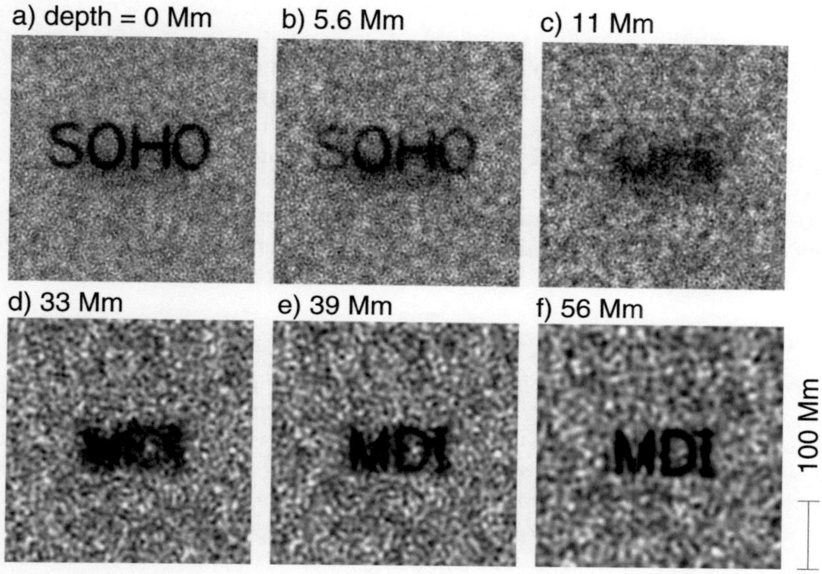

Fig. 4. Egression-power maps of artificial seismic noise that encounters alphanumeric absorbers just beneath the surface and at a depth of 56 Mm. In this simulation, reproduced from Lindsey & Braun (2000b), the absorbers are confined to infinitely thin sheets. Submergence of the focal plane beneath an absorber results not in its disappearance but rather a defocusing of the signature. Diffuse signatures beneath the surface absorbers (frames b and c) and above the submerged absorbers (frames d and e) are a defocus artifact of absorption only at the surface or 56 Mm beneath it and not a signature of additional absorption between those depths.

It is important to observe the distinction between egression-power signatures and the physical properties of the medim that give rise to them. The egression-power signatures of sharp or compact features in egression power maps, for example, may look rather like the source distributions that generate the acoustic waves the signatures represent. However, a diffuse signature may simply represent a source some distance from the focal plane, which itself contains no sources or absorbers whatever. Skartlien (2001) and Skartlien (2002) develop the interesting problem of focus-defocus diagnostics to recover realistic source distributions from egression-power signatures.

2. Helioseismic observations

A wide variety of helioseismic observations have been developed since the advent of helioseismology. The best of these in terms of spatial resolution and stability are from space-borne observatories, the Michelson-Doppler Image (MDI) (Scherrer et al. , 1995) aboard the Solar and Heliospheric Observatory (SOHO), launched by NASA[2] in 1996 and operating into late 2010, and the Helioseismic-Magnetic Imager (HMI) aboard the Solar Dynamics Observatory (SDO), launched early in 2010, also by NASA. The space-borne observatories have been reinforced by observations from a world-wide network of ground-based helioseismic observatories, the Global Oscillations Network Group (GONG), with headquarters at the National Solar Observatory (NSO), in Tucson, Arizona.

During its term of operation, SOHO/MDI made full-disk line-of-sight Doppler maps at a cadence of 60 s with a spatial sampling of $\sim$2 arcsec in its medium-resolution mode of operation, but had a high-resolution, $\sim$0.5 arcsec, mode that covered only part of the solar disk. The SOHO/MDI Dopplergrams were supplemented by occasional intensity and line-of-sight magnetic observations. Since about the turn of the century, the GONG has made full-disk line-of-sight Doppler and magnetic maps and intensity maps, with a resolution of $\sim$2.5 arcsec at a cadence of 60 s.

The SDO/HMI instrument now makes full-disk Doppler, intensity and line-of-sight magnetic maps from a variety of filtergrams at an effective cadence of 45 s with a resolution of $\sim$0.5 arcsec. It also makes Stokes magnetic maps with a cadence of 9 min.

3. Acoustic-power holography of magnetic regions

Figure 5 shows the results of subjacent-vantage acoustic-power holography applied to an actual magnetic region. The upper row shows maps of visible intensity (left) and line-of-sight magnetic field (right) from SOHO/MDI. The middle row shows an acoustic-power map (left) and a subjacent-vantage egression-power map (right) focused at the surface of the region, both in the 5 mHz spectrum (200-s period). The acoustic-power map shows that surface acoustic motion is heavily suppressed in the magnetic regions (Braun et al. , 1992). The egression-power map shows that acoustic radiation emanating downward from the magnetic regions is similarly suppressed.[3] This phenomenon was discovered a decade before the first applications of helioseismic holography by Braun, Duvall & LaBonte (1988) by discriminating ingoing and outgoing wave fluxes in annuli surrounding sunspots. This was a major impetus in the recognition of local helioseismology as a major new field in solar seismology that would include the practical development of helioseismic holography.

Both the acoustic-power maps and the egression-power maps show greater-than-normal power in the peripheries of the active region, where the magnetic field is relatively weak but still greater than it generally is in the quiet Sun. However, the relative distributions of power in "acoustic-power halos" and "acoustic-emission halos" are significantly different. For acoustic radiation with skip distances in the range 15–45 Mm, excess acoustic emission tends to be conspicuous in "magnetic neutral lines", loci along which the magnetic field is approximately horizontal, hence the vertical component of it vanishes (i.e., the line-of-sight component when

---

[2] National Aeronautics and Space Administration

[3] Because the Doppler signature, ψ, is strongly suppressed in the magnetic region, the power of the raw egression, computed as prescribed by equation (8) is artificially suppressed when the pupil is contaminated by magnetic fields. This suppression is roughly corrected in the middle-right panel by dividing the egression power by a smeared version of the power of the "coherent acoustic *in*gression", a time-reverse analogy of the acoustic *e*gression we will introduce formally in §4.1.

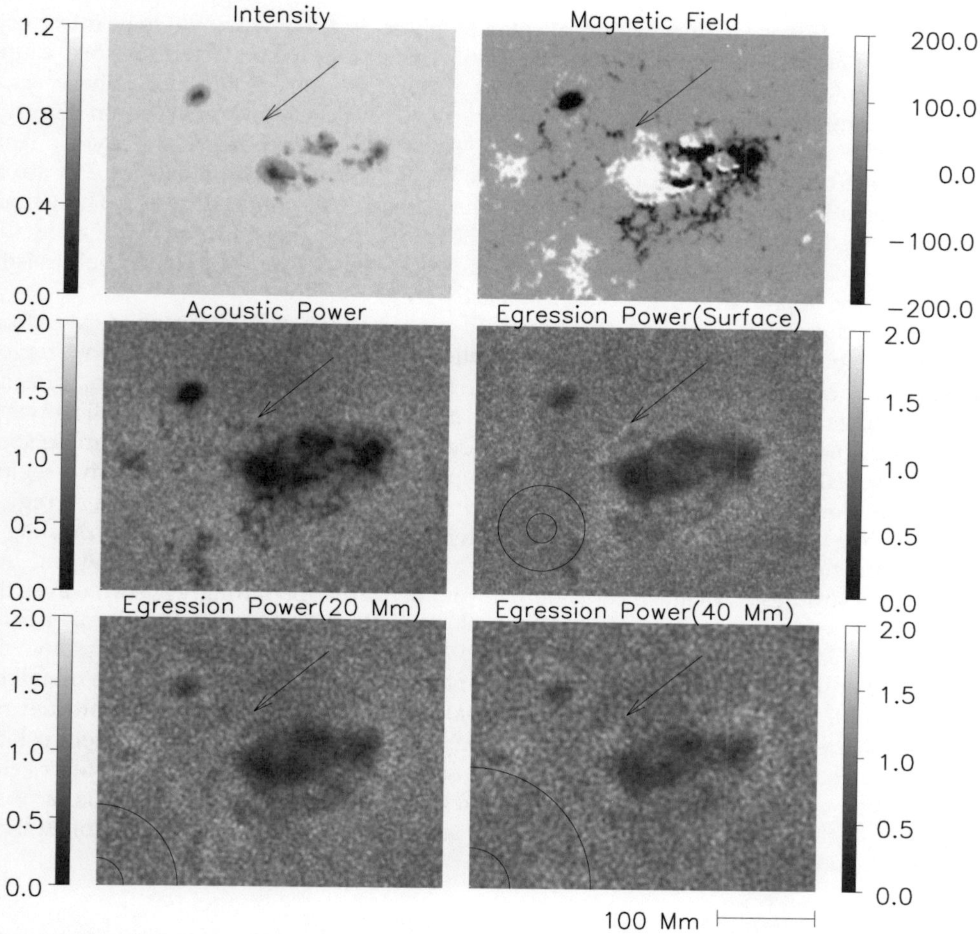

Fig. 5. Egression-power maps of NOAA Active Region 8179. Top panels show intensity (left) and line-of-sight magnetic-field (right) snapshots from SOHO/MDI. Middle-left panel shows a plain 5-mHz acoustic-power map of the region. Middle-right panel shows an egression-power map of the same focused at the surface. The pupil is an annular region with dimensions shown in the lower-left of the panel centered on the focus. Bottom-left panel shows an egression-power map focused at a depth of 20 Mm. Bottom-right panel shows an egression-power map focused at a depth of 40 Mm. An arrow, reproduced in all six frames, points to an excess of seismic emission in the middle-right frame. Acoustic-power and egression-power maps are integrated over a 1 mHz spectrum centered at 5 mHz (i.e., a period of 200 s) and over a duration of 24 hours beginning at 1998 March 15 11:00 UTC.

the active regions is near disk center). An arrow reproduced in all six frames locates such a feature.

The bottom row shows egression-power maps focused 20 and 40 Mm beneath the surface of the active region. The pupils are expanded with increasing depth to keep the vantage subjacent. Because of the increased sound speed, the wavelength of 5-mHz acoustic radiation is greater at greater depths, hence the effects of diffraction are coarser. The main effect of the submerged focal plane appears to be smearing of the egression-power signature.

Interpretation of these signatures is complicated by phase errors now known to be introduced by magnetic fields in the pupils of the acoustic regressions when active regions cannot be entirely avoided (Lindsey & Braun , 2005a). Efforts to account for these phase errors (Lindsey & Braun , 2005a) indicate that surface features such as those seen in Figure 5 are mostly the signature of acoustic anomalies in a relatively thin surface layer. Some investigators (Kosovichev, Duvall & Scherrer , 2000) have suggested the existence of strong acoustic anomalies extending 10 or more Mm beneath sunspot photospheres. The general consensus based on helioseismic holography, however, has been that acoustic anomalies beneath about 2–4 Mm contribute relatively little, perhaps insignificantly, to helioseismic signatures in the neighborhoods of individual active regions.
Because of this, the use of helioseismic holography has developed a strong focus on diagnostics of the relatively shallow subphotospheres of active regions (Lindsey, Cally & Rempel , 2011; Moradi et al. , 2010) as well as the quiet Sun. Results of these studies over the approximately 14 years since the first practical applications of helioseismic holography are considerable, and their descriptions would require more space than we can realistically appropriate in this chapter. Seismic holography of active regions is giving us deep insight into the role of flows in the neighborhoods of active regions (Braun, Birch & Lindsey , 2001), as well the physics of slow- and fast-mode coupling of magneto-acoustic waves in active regions, which seems to be at the heart of how and why active regions suppress acoustic motion in their photospheres and strongly absorb waves that the quiet photosphere normally reflects (Cally , 2000; Cally & Bogdan , 1997; Schunker et al. , 2008; Spruit & Bogdan , 1992). The preponderance of these results involves physics somewhat beyond the familiar scope of optics and holography. We will therefore devote the remainder of this article to two particular phenomena that appear to most easily illustrate the role optics has taken on in helioseismology: (1) acoustic-power holography of transient seismic emission from solar flares, and (2) the use of phase-correlation holography to monitor active regions in the Sun's far hemisphere. Excellent reviews of scientific results from solar acoustic holography in the general context of local helioseismology are contained in Gizon & Birch (2005) and Gizon et al. (2009).

4. Holography of seismic transient emission from flares

4.1 Egression-power signature of a flare

In the solar flare of 1996 July 09, Kosovichev & Zharkova (1998) discovered the first instance of an acoustic transient released into the solar interior by a flare. The surface manifestation of this was a pattern ripples seen in Doppler maps of the active region propagating away from the site of the flare from 15 minutes to an hour after the impulsive phase, what Kosovichev & Zharkova (1998) called a "sunquake". Sunquakes are the surface signature of waves that have traveled tens of thousands of km beneath the foot points of the flare and come back to the surface tens of thousands of km horizontally from the same. Figure 6 shows this phenomenon in a flare in which it was most conspicuous, the X1.2-class flare of 2005 January 15 from NOAA [4] Active Region (AR) 10720. The top row shows intensity (left) and line-of-sight magnetic (right) maps of NOAA AR10720 within an hour before the flare. The middle-left panel shows the Doppler disturbance marking the impulsive phase of the flare, a predominantly red-shifted feature consistent with a downward motion of the photosphere of a few hundred m s^{-1} in a narrow channel aligned along the magnetic neutral line. An arrow, cospatially reproduced in all of the frames, points to the location of this feature. The

[4] National Oceanic and Atmospheric Administration, US Department of Commerce

middle-right panel shows a map of the Doppler signature filtered for temporal variations in a 2-mHz interval centered at 6 mHz 24 minutes after the impulsive phase. A second arrow in this frame points to the most conspicuous surface ripples, 15 Mm above the site of the impulsive disturbance. Nearly cospatial with the impulsive Doppler disturbance shown in the middle-left frame is an impulsive increase in the intensity of a few percent, characteristic of a small white-light flare. This is shown in the lower-left frame.

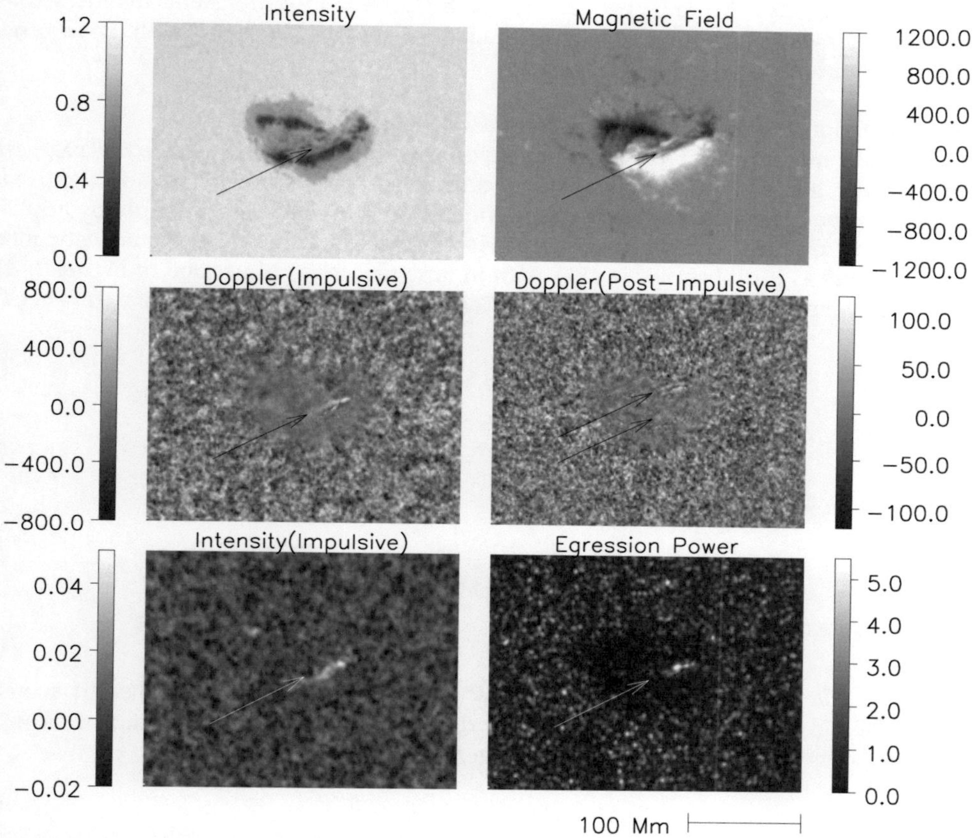

Fig. 6. Helioseismic signature of the X1 flare of 2005 January 15, adapted from Donea (2011). Upper-left panel shows an MDI intensity map of NOAA AR10720 shortly before flare onset. Upper-right panel shows a pre-flare MDI line-of-sight magnetogram of the same. Middle-left panel shows an MDI Doppler map at flare onset, with arrow, reproduced in all other frames, pointing to the sudden, compact red-shift signature at the acoustic source. Middle-right panel shows 6-mHz Doppler signature, ψ, 24 minutes after the onset of the flare. The top arrow in this frame points to surface Doppler ripples proceeding outward from the site located by the lower arrow. Lower-left panel shows the signature of sudden visible continuum emission observed by GONG. Lower-right panel shows an egression-power map of the region during the impulsive phase.

Subjacent-vantage acoustic-power holography of the ripples propagating outward from the site of the impulsive signatures render the source-power distribution of the waves represented by the surface ripples. The lower-right frame shows a map of the 6-mHz egression power extrapolated subjacently backwards from the post-impulsive ripples to the impulsive phase of the flare. This source pattern is seen to conform closely to the site of both the impulsive intensity and the Doppler transient. In some sunquakes, the acoustic source distribution has coincided with apparent magnetic transients. These diagnostics have motivated models in which seismic transients are thermally driven in some instances, by pressure perturbations associated with impulsive heating (Donea & Lindsey , 2005; Kosovichev & Zharkova , 1998; Lindsey & Donea , 2008; Moradi et al. , 2007), and by Lorentz-force transients in others, due to the sudden release of magnetic free energy (Hudson, Fisher and Welsch , 2008). For a general review, we refer to Donea (2011).

4.2 Discrimination of acoustic radiation from a localized source

In an acoustic medium that conforms to invariance under time reversal, the regression operator, $P_+(z)$, that extrapolates the surface acoustic field backwards in time to a depth z has a counterpart that extrapolates it forward in time. This is accomplished simply by applying the surface disturbance, ψ, represented by surface ripples in Figure 1, to the acoustic model forward in time, as it is observed, rather than in time reverse as illustrated in Figure 2. This computation simply replaces the Green's function, G_+, in equation (8) by its time reverse, G_-, hence,

$$H_-(\mathbf{r}, t) \equiv \int_{-\infty}^{\infty} dt' \int_{\mathcal{P}} d^2\mathbf{r}' \, G_-(\mathbf{r}, \mathbf{r}'; t - t') \, \psi(\mathbf{r}', t'), \tag{13}$$

where

$$G_-(\mathbf{r}, \mathbf{r}'; \tau) \equiv G_+(\mathbf{r}, \mathbf{r}'; -\tau). \tag{14}$$

In the frequency domain,

$$\hat{H}_-(\mathbf{r}, \omega) = \int_{\mathcal{P}} d^2\mathbf{r}' \, \hat{G}_-(\mathbf{r}, \mathbf{r}'; \omega)\hat{\psi}(\mathbf{r}', \omega), \tag{15}$$

and in this context,

$$\hat{G}_-(\mathbf{r}, \mathbf{r}', \omega) = \hat{G}_+^*(\mathbf{r}, \mathbf{r}', \omega). \tag{16}$$

We call H_-, the counterpart of the coherent acoustic egression, the "(coherent acoustic) *in*gression", and the operator, $P_-(z)$, that derives it from ψ the "(coherent acoustic) *pro*gression" operator, whereby equation (12) generalizes to,

$$H_\pm(\mathbf{r}, z, t) = P_\pm(z)\psi(\mathbf{r}, 0, t). \tag{17}$$

For a lossless medium that conforms to time-reversal invariance, and in the limit of a pupil, $\mathcal{P}$, with an infinitesimal inner radius that otherwise covers the entire solar surface, the successive application of $P_-(0)$ and $P_+(0)$ returns ψ itself:

$$P_-(0)P_+(0)\psi(\mathbf{r}, 0, t) = \psi(\mathbf{r}, 0, t), \tag{18}$$

so long as ψ represents acoustic waves that propagate downward some distance from the surface into the solar interior before they are turned back to the Sun's surface. This, along with the spatial discrimination the regression gives us with respect to waves emitted from the source region, makes it possible to isolate the component of acoustic radiation emanating from just the source region. Figure 7 demonstrates this exercise applied to the flare of 2005 January 15. The top two frames show the 6-mHz Doppler disturbances, ψ, during the impulsive phase (left) and 24 minutes after (right). Having applied $P_+(0)$ to ψ we examine the distribution in egression-power, $|H_+(\mathbf{r},t)|^2$, during the impulsive phase and outline the source region of interest. This is rendered by the parameceoid inset of the egression-power distribution in the middle-right frame of Figure 7. We then apply a spatial mask, M, admitting only this part of $H_+(\mathbf{r}, t)$, and finally apply $P_-(0)$ to the masked egression to reverse the regression. The full operation, then, is

$$\psi_M(\mathbf{r}, 0, t) = P_-(0)MP_+(0)\psi(\mathbf{r}, 0, t). \tag{19}$$

The result, ψ_M, 24 minutes after the impulsive phase is rendered in the middle-right frame of Figure 7 surrounding of the egression-power inset. The acoustic power, $|\psi_M(\mathbf{r}, t)|^2$, from the source region is rendered directly below, in the lower-right frame. The lower-left frame plots the power at this moment along a circle of radius 22.4 Mm, drawn in the lower-right frame. The non-uniform directional distribution of the acoustic radiation is the signature of a source that is not only spatially extended, as is evident from the egression-power map, but one that is undergoing complex motion (Donea & Lindsey , 2005; Kosovichev , 2007).
These diagnostics offer considerable promise towards an understanding of the nature of the impetus that drives transient seismic emission into the solar interior.

5. Seismic signatures of active regions in the Sun's far hemisphere

Time-distance statistics by Duvall et al. (1993) showed that acoustic radiation in the 2.5–4.5 mHz band propagates from the near hemisphere to the far hemisphere and back, retaining significant coherence. It might seem, then, that seismic imaging of the Sun's far hemisphere would be simply a matter of extending the pupil, $\mathcal{P}$, of a holographic regression to the hemisphere opposite to the focus. The understanding here is that the pupil is in the hemisphere accessible to observations of the acoustic field, ψ; hence, the focus is in the far hemisphere. It turns out that the preponderance of acoustic radiation that survives the trip from one hemisphere to the other does so by benefit of at least one specular reflection from the surface.[5] But, these species of acoustic radiation, those which penetrate deep into the Sun's interior and skip long distances, appear to be rather poorly absorbed by most of the magnetic region. For these waves, magnetic regions act more as scatterers than absorbers. Because of this, seismic monitoring of the Sun's far hemisphere is based heavily on two basic elements:

1. phase-correlation seismology, and
2. the extension thereof to multiple-skip acoustics.

5.1 Phase-correlation holography

Acoustic-power holography is highly effective for detecting and locating local absorbers or emitters in an environment in which these are prevalent. However, it is not generally very effective for detecting elastic scatterers in a non-absorbing acoustic medium bounded by a specularly reflecting surface. A scatterer does indeed block radiation that would otherwise have registered at the focus of a holographic computation, casting a silhouette with respect to it in technical terms. However, in a nominally isotropic radiative environment, the scatterer generally replaces the radiation it blocks with radiation that would have missed the focus, filling the silhouette with radiation as bright as that which was blocked. The imposition upon efforts to detect a scatterer based on egression power, then, is rather like that upon having to see white cat in a white room.[6] It is rather the character of an efficient scatterer to shift the *phase* of radiation that encounters it than its intensity.

[5] Duvall et al. (1993) found that the Sun's surface becomes a strong absorber of acoustic radiation much above 4.5 mHz in frequency, a result Lindsey & Braun (1999) confirmed by 2-skip seismic holography (see also Braun & Lindsey , 2000a).

[6] We thank former NSO summer student Mark Fagan for this metaphor.

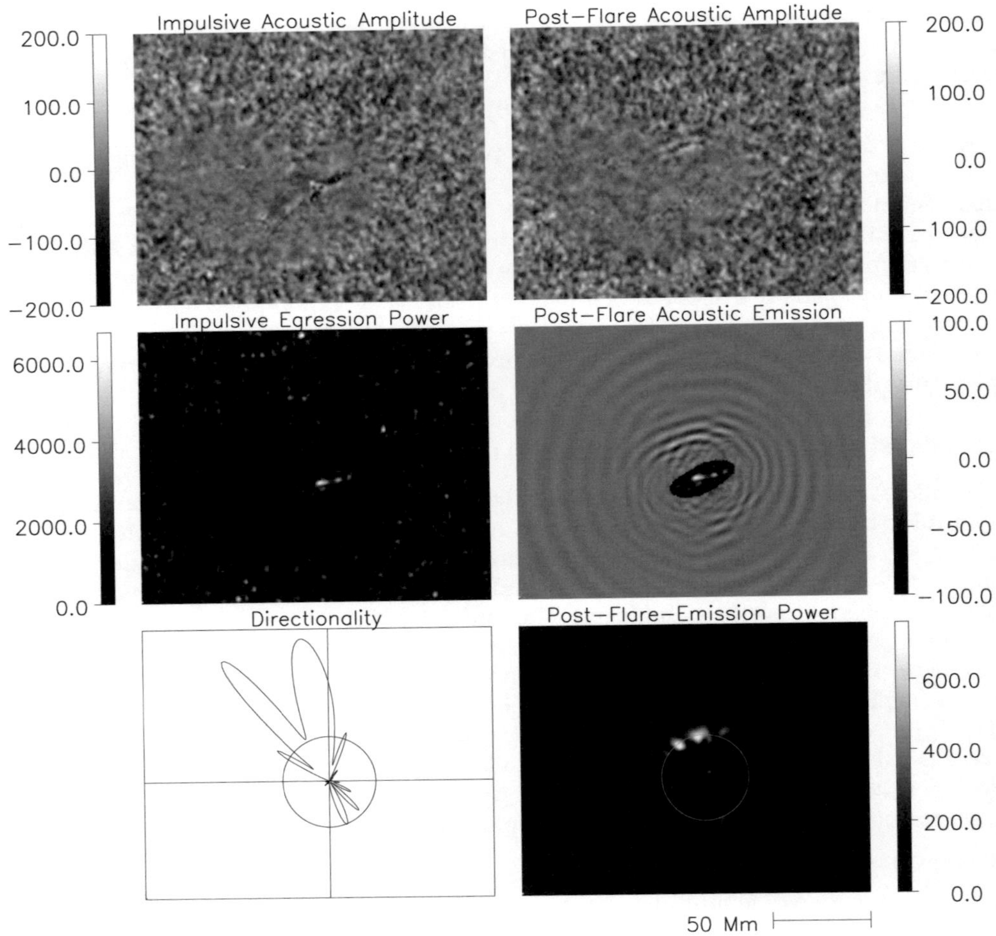

Fig. 7. Discrimination of the component of the 6-mHz Doppler signature, ψ, emanating from the site of transient emission shown in Figure 6, adapted from Donea (2011). Upper-left panel shows the 6-mHz Doppler signature, ψ, at flare onset. Upper-right panel shows the same 24 minutes after flare onset. Middle-left panel shows the 6-mHz egression power at flare onset. Middle-right panel shows a holographic representation of the component, ψ_M, of ψ that has emanated from the source region, marked by the inset of the egression power from the source region, 24 min after flare onset. Lower-right panel shows the acoustic power, $|\psi_M|^2$, emanating from the source region 24 min after flare onset. Lower-left panel plots the acoustic power, $|\psi_M|^2$, along the 22.4-Mm-radius circle in the lower-right frame. The circle plotted in the lower-left frame represents an egression power of 200 m^2 s^{-2}.

Figure 8 shows 5-mHz phase maps of cospatial correlations between H_+, H_-, focused subjacently at the Sun's surface from the near hemisphere, and with ψ, in the neighborhood of NOAA AR8179. H_+ and H_- are computed here over the same pupil as applied in the middle right frame of Figure 5. Top two frames show intensity (left) and line-of-sight magnetic field (right). Middle left frame shows normalized egression-power, $\langle|H_+(\mathbf{r},t)|^2\rangle$, reproduced from Figure 5; and the arguments of $\langle H_+H_-^*\rangle$ (middle-right); $\langle H_+\psi^*\rangle$ (lower-left);

and $\langle \psi H_-^* \rangle$ (lower-right). We call the correlations of $H_\pm$ with ψ (bottom row) the "control correlations", since they compare extrapolations of ψ from the pupil with ψ itself. The "ingression control correlation" phase, arg $\langle \psi H_-^* \rangle$, in an active region expresses how the active region shifts the phases of the acoustic signatures of waves impinging into it from beneath the photosphere. The egression-ingression correlation $\langle H_+ H_-^* \rangle$ characterizes how the active region reflects these same waves back into the Sun's interior. The negative deflection in its phase, arg $\langle H_+ H_-^* \rangle$, for foci in the active region is the signature of reflected waves that arrive into the egression pupil up to ~100 s ahead of their counterparts reflecting from the quiet solar surface. In recent models of sunspots, these reduced travel times appear to be largely the result of something like a ~300 km depression in the photospheres of sunspot umbrae (Lindsey, Cally & Rempel , 2011). This interpretation is a great oversimplification of the reality. However, because it works very well as a model, we have characterized this seismic quality of active regions as an "acoustic Wilson depression", bearing in mind a likely relationship to the well-known Wilson effect discovered by Alexander Wilson in the 16th century. In the case of the signatures shown in Figure 8, the phase shifts tend to be enhanced in sunspot penumbrae, where the magnetic field is highly inclined. However, these effects extend far outside of the sunspots into regions in which the magnetic field is less than 1 kG. Because of this, the integrated seismic signature of the active region is several times that due to the sunspots alone. This greatly enhances the acoustic visibility of a large active region in the Sun's far hemisphere. And, unlike in the case of absorption, the phases of waves that skip long distances are roughly as sensitive sensitive to the acoustic Wilson depression as those whose phase perturbations are mapped in Figure 8.

5.2 Multiple-skip holography

Because the sound speed in the solar interior increases by such a large factor from the surface to the Sun's core, only a small fraction of the acoustic radiation generated at the surface penetrates deep beneath the Sun's surface before coming back to the surface. The ray paths that connect a point at the center of the far hemisphere to a pupil in the near hemisphere whose radius is 0.9 times the radius of the solar disk must emanate downward within a vertical cone whose half-angle, α, is 0.22°, a rapidly decaying quasi-exponential function of the skip distance. The resulting diffraction limit,

$$w = cT/\sin\alpha, \tag{20}$$

—where $c = 8$ km s^{-1} is the photospheric sound speed and $T \sim 300$ s, is the wave period—is prohibitive: For a conical half-angle, α, of 0.22°, w is 630 Mm, nearly a solar radius and much larger than the largest active regions, even including their plages.

Fortunately, waves with periods of ~300 s incident into the Sun's surface from beneath it are known to be largely reflected back into the solar interior Duvall et al. (1993). The reflection is of good specular quality, such that coherence is significantly preserved over a several skips beneath the Sun's surface (Lindsey & Braun , 1999). It is straight-forward to formulate the Green's functions, $G_\pm$, to account for this, making multiple-skip holography practical. Ray paths that cover the same distance in two skips, by reflecting once from the solar surface subtend a cone that is much wider (i.e., not nearly as decayed), by a factor of a factor of about five, to 1.2°. The extra skip, then, greatly improves the diffraction limit, w, to ~120 Mm, the scale of a moderately large active region.

Figure 9 shows a diagram of ray paths and matching wave-front geometries representing the basic acoustical elements of 2×2-skip phase-correlation holography of the Sun's far hemisphere (top) from the vantage of the near hemisphere (bottom), as derived by the far-side

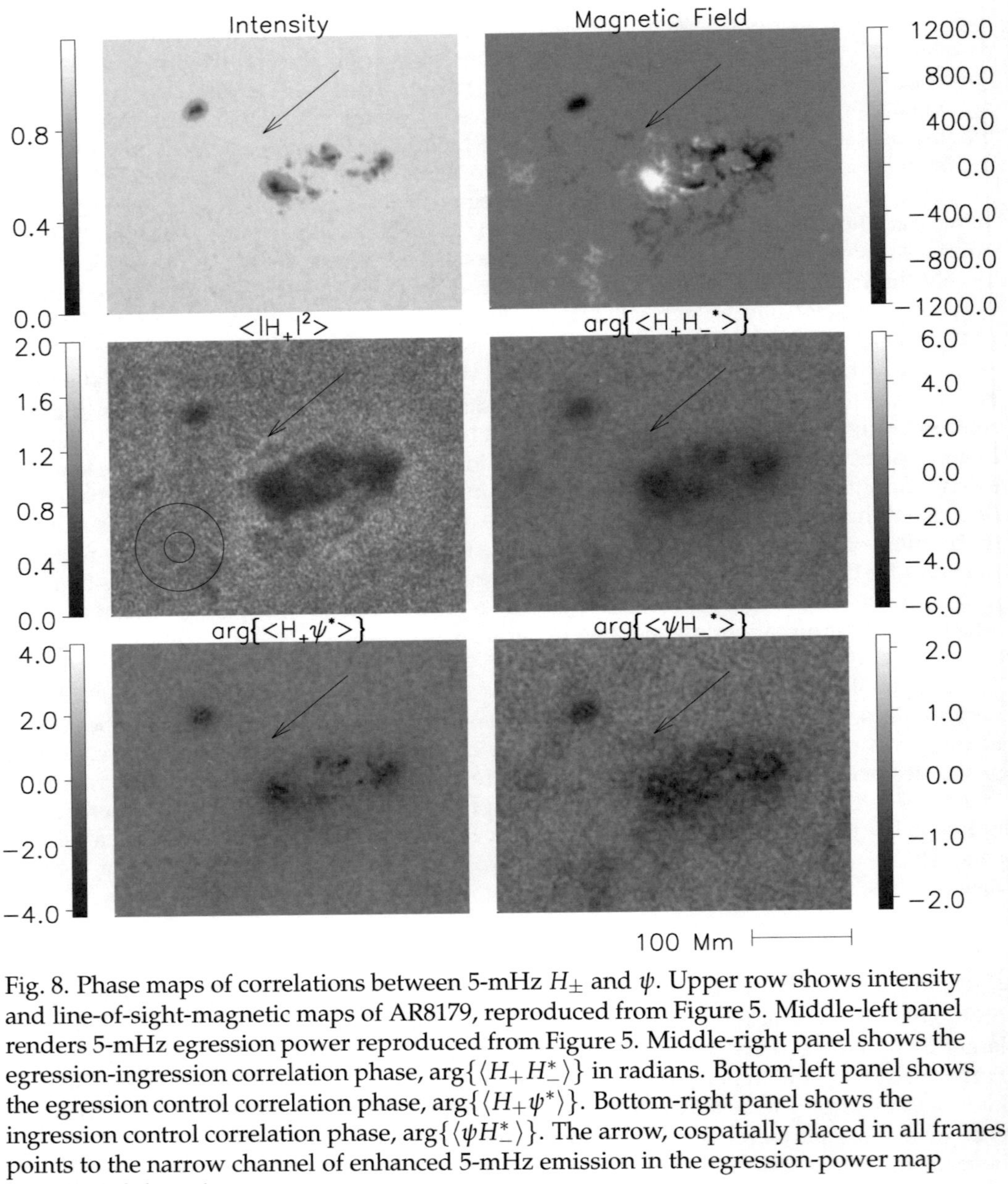

Fig. 8. Phase maps of correlations between 5-mHz $H_{\pm}$ and ψ. Upper row shows intensity and line-of-sight-magnetic maps of AR8179, reproduced from Figure 5. Middle-left panel renders 5-mHz egression power reproduced from Figure 5. Middle-right panel shows the egression-ingression correlation phase, $\arg\{\langle H_+ H_-^* \rangle\}$ in radians. Bottom-left panel shows the egression control correlation phase, $\arg\{\langle H_+ \psi^* \rangle\}$. Bottom-right panel shows the ingression control correlation phase, $\arg\{\langle \psi H_-^* \rangle\}$. The arrow, cospatially placed in all frames points to the narrow channel of enhanced 5-mHz emission in the egression-power map (middle-left frame).

seismic synoptic monitors presently operating at the headquarters of the NSO/GONG project, in Tucson, and the Joint Science Operations Center of the SDO, at Stanford University. The acoustic travel times along the trajectories plotted in Figure 9 are approximately 3.5 hours from the near hemisphere to the focus in the far hemisphere. The round trip travel time, then, for a disturbance in the near hemisphere to its echo in the near hemisphere from the focus is about 7 hours. To detect the phase correlation between H_+ and H_-, then, requires observations of the near hemisphere for at least this period before correlation statistics can begin to accumulate.

The first far-side seismic maps, based on 2×2-skip acoustics were published by Lindsey & Braun (2000a), computed from helioseismic observations by SOHO/MDI. The Solar Oscillations Investigation (SOI) at Stanford University implemented a synoptic far-side seismic monitor using the 2×2-skip algorithm in early 2001, which continued to operate until the recent expiration of SOHO/MDI in early 2011.

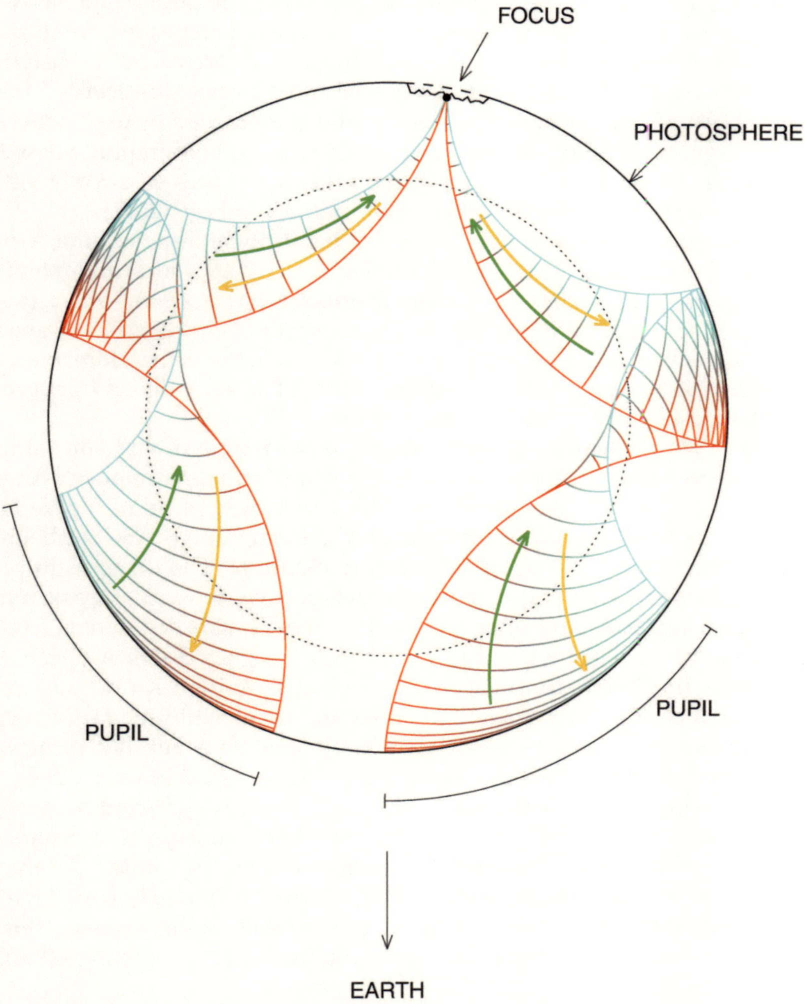

Fig. 9. Diagram of ray paths and matching wave-front geometries representing 2×2-skip phase-correlation holography of the far (top) hemisphere of the Sun from observations over a pupil in the near (bottom) hemisphere.

The 2×2-skip algorithm is only sensitive to active regions within ∼50° of the antipode of disk center in the near hemisphere. Braun & Lindsey (2001) extended the algorithm to cover the full far hemisphere by incorporating 1×3-skip acoustics. Both the GONG and the SDO now have synoptic far-side seismic monitors covering the full far hemisphere of the Sun.

Figure 10 shows maps of the phase of $\langle H_+ H_-^* \rangle$ of the far hemisphere computed from GONG observations, using the foregoing schemes to cover the full far hemisphere. In this presentation, the "Carrington" mapping, the Sun's surface is rendered by longitude and latitude in a co-rotating reference frame in which active regions are nearly stationary. Hence, it is the region viewable from Earth that moves, from right to left as the Sun rotates from east (left) to west (right) with respect to Earth. Each map represents a compilation of statistics from observations over a 24-hour period, effectively 17 hours when an account is taken of the 7-hr round-trip travel time. The signature of what was to be designated NOAA AR10808 is seen passing across the far eastern hemisphere from Earth perspective, rotating into direct view a week after crossing far-side central meridian. Upon its arrival at the eastern limb, this active region released an X17-class flare, one of the most intense in recorded history—and one of the best observed, largely because of preparations motivated by expectations based on the holographic signatures. For a deeper discussion relating holographic phase-correlation signatures in the far hemisphere to magnetic and other signatures as directly viewed in the near hemisphere, we refer to González Hernández, Hill & Lindsey (2007).

The ability to monitor activity in the Sun's far hemisphere has become a major asset in space-weather forecasting (González Hernández et al. , 2009), including solar-irradiance forecasting (Fontenla, et al. , 2009). UV solar irradiance has a strong effect the terrestrial ionosphere and exosphere, which determines the rates at which the orbits of spacecraft and space debris decay. Improved forecasting of solar UV irradiance is therefore a major object of agencies whose task is to keep track of a large inventory of orbiting debris from ground-based observations that can be interrupted by poor weather.

It also occasionally happens that a flare or coronal mass ejection (CME) emanating from the Sun's far hemisphere has a significant effect on the near-Earth environment. An example of this was the halo CME of 2001 August 15–16. The upper-right panel of Figure 11 shows an image of the CME in the Large Angle Spectrometric Coronagraph (LASCO) aboard SOHO at the outset of its progression into the interplanetary medium. The lower-right panel shows the 2×2-skip phase-correlation map from Earth perspective, the signature of a newly born active region appearing below and somewhat to the left of far-side disk center, having crossed far-side central meridian about a day before. NOAA designated this AR09591 a few days after it rotated into direct view from Earth. The lack of significant X-ray emission from this event (lower-left panel) is simply a result of its having occurred in the far hemisphere. I.e., copious X-rays must certainly have been released by such an event, but these would have been radiated into the far side of the solar system, hence invisible from Earth. This does not apply to high-energy charged particles, which are strongly deflected by magnetic fields that are probably involved in their acceleration, possibly a considerable distance from the location of the active region as indicated by its helioseismic signature. As the upper-left panel of Figure 11 shows, a considerable flux of high-energy protons showered the near-Earth environment promptly following the appearance of this CME. In this instance, this happened to be of significant concern to crew members of the International Space Station (ISS), who were undertaking an extra-vehicular activity (EVA) at the time.

Events such as that shown in Figure 11 happen rarely, perhaps only once in an 11-year solar-activity cycle, and helioseismic signatures of active regions in the far hemisphere cannot give us a very reliable assessment of the potential of an active region in the far hemisphere breaking all rules of fairness and decency to impose upon the near-Earth environment this way. Indeed, we do not yet have this capability even for an active region in direct view, in which case the particle flux to be expected is many times greater, posing a significant health hazard to an exposed crew. The ability to monitor large active regions in the Sun's far hemisphere nevertheless greatly facilitates planning of activities to which flares pose a

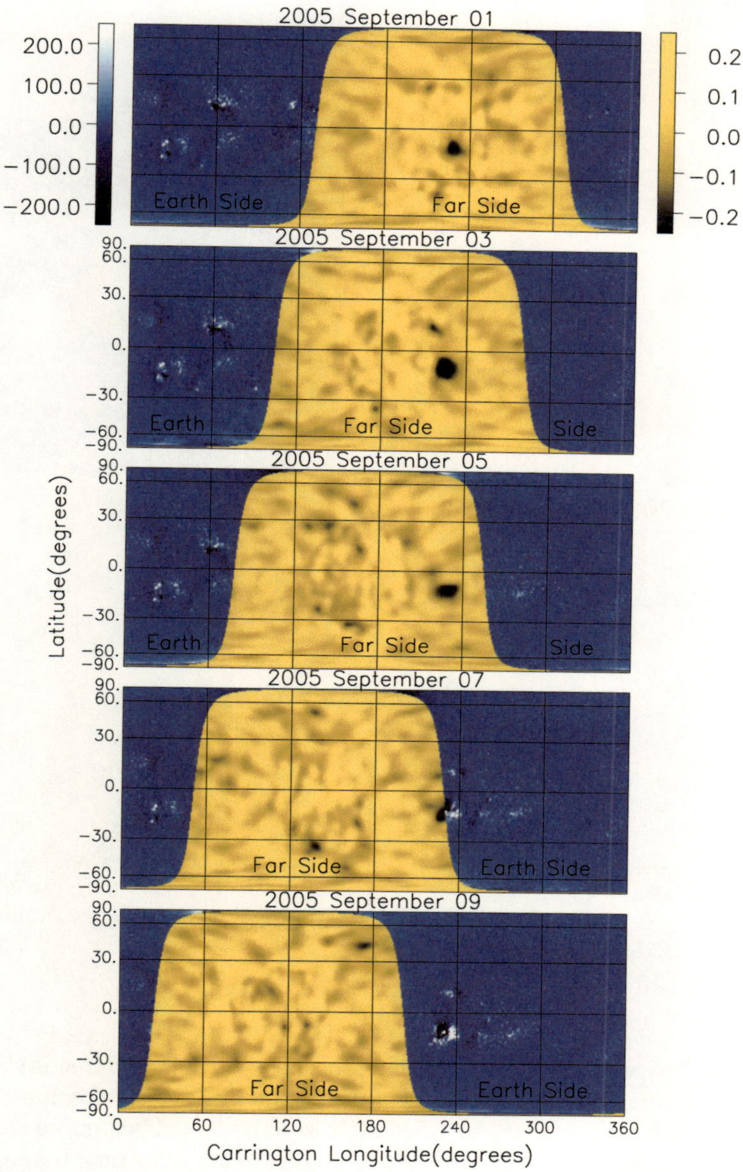

Fig. 10. The phase of $\langle H_+ H_-^* \rangle$ focused on the surface of the Sun's far hemisphere is compared with line-of-sight magnetic maps in the near hemisphere as viewed directly from earth over the period 2005 September 01–09. The signature of NOAA AR10808 is seen just after crossing far-side central meridian on September 01, the region rotating into direct view from Earth on September 07.

considerable liability, particularly on time scales of a week or two when the existences of active regions that could produce them have yet to be announced by any other means.
Since its implementation in the early 2000s, the helioseismic monitor of the Sun's far hemisphere has been complemented by measurements of Ly-α radiation back-scattered from

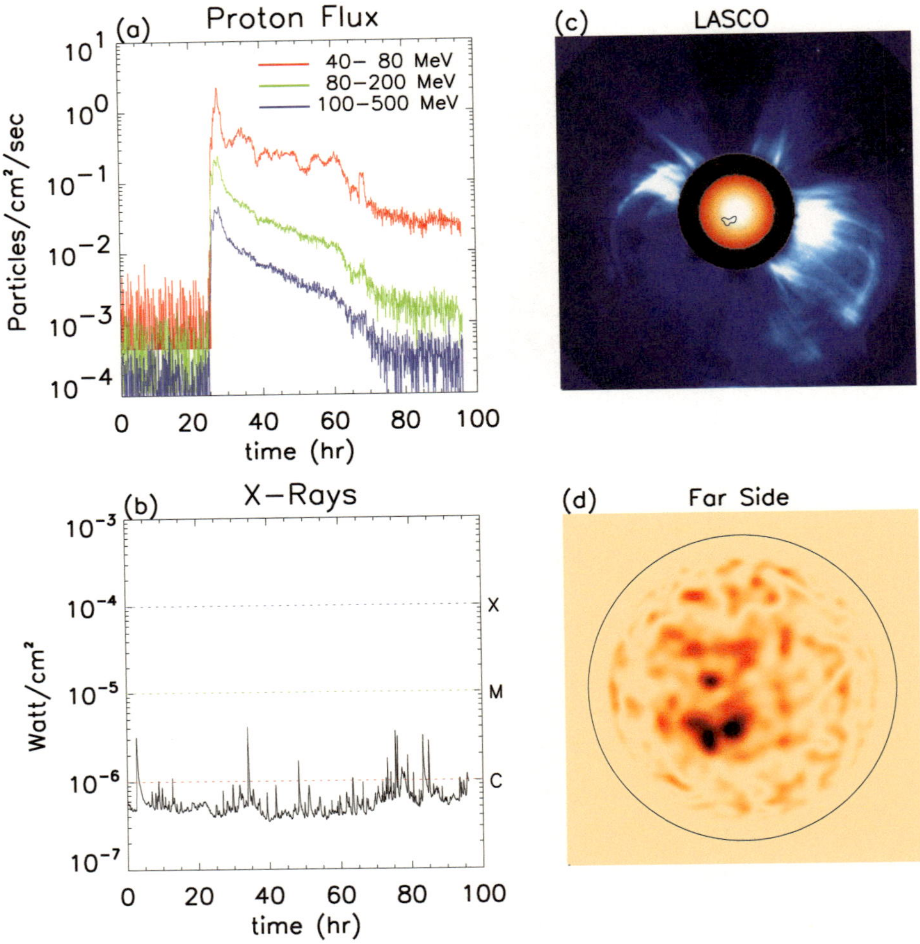

Fig. 11. High-energy-proton and X-ray fluxes associated with a large far-side CME occurring on 2001 August 15–16 . Panel *a* shows plots of the proton flux detected by the GOES–10 spacecraft at energies ranging from 40 to 500 MeV in 96-hr period beginning at 2001 August 15, 00:00 UT. Panel *b* shows plots of the 1–8Å X-ray flux in the same time frame. Panel *c* shows a LASCO image of the CME at August 16, 00:31 UT, approximately one hour after first evidence of the CME in the LASCO coronagraphs. The Sun's surface is represented by a solar icon at the center of the LASCO image. A closed contour drawn on the solar icon represents the location of active regions in the Sun's far hemisphere that were to be designated 9557 and 9591 a few days later when they rotated into direct view. Panel *d* shows a far-hemisphere map of arg $\langle H_+ H_-^* \rangle$ as viewed from LASCO perspective through the near hemisphere and intervening solar interior. The composite signature of the two active regions is seen approximately 0.3 solar radii south (below) and slightly east (left) of far-side disk center.

the interplanetary medium in the far side of the solar system, the major component of which originates from active regions in the Sun's far hemisphere (Fontenla, et al. , 2009; Quemerais & Bertaux , 2006). The instrument that makes these measurements is the Solar Wind ANisotropies (SWAN) experiment, on SOHO.
As of early 2011, NASA's two STEREO spacecraft have had full coverage the Sun's far-hemispheric corona with X-ray observations. The STEREO spacecraft will continue to enjoy this full far-hemispheric coverage until about 2019.

6. The future of helioseismic holography

Helioseismic holography will certainly continue to play a major role in helioseismology, and in solar research at large, for the foreseeable future. Nearly all of the applications to date have focused on relatively compact anomalies near the Sun's surface. This is largely because these manifest the strongest and most compact signatures, the ability of which to discriminate spatially is the significant advantage of the optical perspective. And, in certain respects, these superficial features play the role of an "acoustic showerglass" we have to look through to see what lies beneath them (Lindsey & Braun , 2005a;b). However, the availability of ever more powerful computing facilities encourages the extension of seismic holography deep into the solar convection zone. This offers the possibility of new insight into the workings of the solar dynamo, the origin of emerging solar activity.
The prospect of seismic holography of the deep solar interior is further encouraged by plans under development by NASA and ESA[7] to include helioseismometers on spacecraft in heliocentric orbits with direct vantages into the Sun's far hemisphere. Simultaneous seismology of both hemispheres offers the best prospects for rotational diagnostics of the Sun's core, which maintains the nuclear reactions that have kept the Sun alive for eons.
The application of seismic holography to more than a hundred emerging active regions during solar cycle 23 (Birch et al., 2009) shows tantalizing signatures up to three days before the significant emergence. These signatures are not strong enough to forecast active-region emergences on an individual basis. However, the statistical existence of these signatures has major implications respecting the dynamics of magnetic flux approaching the Sun's surface from below. This entails the tantalizing suggestion that such a forecast will be possible once we understand the dynamics underlying the pre-emergence signatures.
In summary, then, this book adds to a wide consensus that optics must certainly be one of the extremely few most powerful diagnostic tools nature has given us. The extension of electromagnetic optics to other wave-mechanical resources has already been a very welcome development such as in electron microscopy, acoustic microscopy, under-water acoustics and geoseismology. It is now beginning to render major benefits in our understanding of the interior workings of the star we live by. Helioseismic holography is a young and growing field of scientific research. We are convinced it will lead to many satisfying benefits in the coming generation.

7. Acknowledgments

We dedicate this chapter to the memory of solar physicist Karen Lorraine Harvey (1942–2002). Many people have contributed to the technical development of solar acoustic holography since it was first proposed, 35 years ago, especially in the past 20 years. We are especially grateful for the unflagging support of Jack Harvey, at the National Solar Observatory, and his late wife, Karen, who supported this work with the resources of the Solar Physics

[7] European Space Agency

Research Corporation. We thank Aaron Birch for his consultation. The development of helioseismic holography has received invaluable support from the Solar-Terrestrial and Astronomy-and-Stellar-Astrophysics Branches of the National Science Foundation and from the National Aeronautics and Space Administration during the past two decades.

8. References

Birch, A. C. & Braun, D. C.; Leka, K. D.; Barnes, G.; Dunn, T. L. & González Hernández, I. (2009). A Search for pre-emergence signatures of active regions. *Bul. Am. Astronom. Soc., Sol. Phys. Div.*, Vol 41, 810

Born, M. & Wolf, E. (1975a). *Principles of Optics*, 491–505, Pergamon Press, Oxford

Born, M. & Wolf, E. (1975b). *Ibid.*, 375–378

Braun, D. C.; Duvall, T. J. Jr. & LaBonte, B. J. (1988). The absorption of high-degree p-mode oscillations in and around sunspots. *Astrophys. J.*, Vol. 335, 1015–1025

Braun, D. C.; Lindsey, C.; Fan, Y. & Jefferies, S. M. (1992). Local acoustic diagnostics of the solar interior. *Astrophys. J.*, Vol. 392, 739–745

Braun, D. C., Lindsey, C., Fan, Y. & Fagan, M. (1998). Helioseismic holography of solar activity. *Astrophys. J.*, Vol. 502, 968–980

Braun, D. C. & Lindsey, C. (2000a). Helioseismic holography of active-region subphotospheres. *Solar Phys.*, Vol. 192, 285–305

Braun, D. C. & Lindsey, C. (2000b). Phase-sensitive holography of solar activity. *Solar Phys.*, Vol. 192, 307–319

Braun, D. C. & Lindsey, C. (2001). Seismic imaging of the far hemisphere of the Sun. *Astrophys. J. Letters*, Vol. 560, 189–192

Braun, D. C.; Birch, A. C. & Lindsey, C. (2004). Local helioseismology of near-surface flows. *Helio- and Asteroseismology: Towards a Golden Future. Proceedings of the SOHO 14/GONG 2004 Workshop*, E. Danesy (Ed.), (ESA SP559) pp. 337–340, ESA, Yale University, New Haven

Christensen-Dalsgaard, J.; Proffitt, C. R. & Thompson, M. J. (1993). A new technique for measuring solar rotation. *Astrophys. J. Letters*, Vol. 403, 75–78

Cally, P. S. (2000). Modelling p-mode interaction with a spreading sunspot field. *Solar Phys.*, Vol. 192, 395–401

Cally, P. S. & Bogdan, T. J. (1997). Simulation of f- and p-mode interactions with a stratified magnetic field concentration. *Astrophys. J. Letters*, Vol. 486, 67–70

Chang, H.-K., Chou, D.-Y., LaBonte, B. J. & the TON Team (1997). Ambient acoustic imaging in helioseismology. *Nature*, Vol. 389, 825–827

Chou, D.-Y. (2000). Acoustic imaging of solar active regions. *Solar Phys.*, Vol. 192, 241–259

van Cittert, P. H. (1939). Kohärenz probleme. *Physica*, Vol. 6, 1129–1138

Claerbout, J. F. (1997). *Imaging the Earth's Interior* Nikos Dracos, Computer Based Learing Unit, University of Leeds

Deubner, F.-L. (1975). Observations of low wavenumber nonradial eigenmodes of the Sun. *Astron. & Astrophys*, Vol. 44, 371–375

Donea, A.-C. (2011). Seismic transients from flares in solar cycle 23. *Space Science Reviews* in press

Donea, A.-C.; Braun, D. C. & Lindsey, C. (1999). Seismic images of a solar flare. *Astrophys. J. Letters*, Vol. 513, 143–146

Donea, A.-C. & Lindsey, C. (2005). Seismic emission from the solar flares of 2003 October 28 and 29. *Astrophys. J.*, Vol. 630, 1168–1183

Donea, A.-C.; Lindsey, C. & Braun (2000a). Stochastic seismic emission from acoustic glories and the quiet Sun. *Solar Phys.*, Vol. 192, 321–333

Duvall, T. L. Jr.; Jefferies, S. M.; Harvey, J. W. & Pomerantz, M. (1993). Time-distance helioseismology. *Nature*, Vol. 362, 430–432

Duvall, T. L. Jr.; D'Silva, S.; Jefferies, S. M.; Harvey, J. W. & Schou, J. (1992). Downflows under sunspots detected by helioseismic tomography. *Nature*, Vol. 379, 235–237

Evans, J. W.; Michard, R.; Servajean, R. (1963). Observational Study of macroscopic inhomogeneities in the solar atmosphere. V. Statistical Study of the time variations of solar inhomogeneities. *Astron. & Astrophys.*, Vol. 26, 368–382

Fontenla, J. M.; González Hernández, I.; Quémerais, E. & Lindsey, C. (2009). Solar irradiance forecast and far-side imaging. *Adv. in Space Sci.*, Vol. 44, 457–464

Gizon, L. & Birch, A. C. (2005). Local Helioseismology. In *Living Reviews in Solar Physics*, S. Solanki, J. Christensen-Dalsgaard, B. Foleck, E. Marsch, R. Rosner, T. Sakurai, K. Schrijver, M. Schüssler & R. Schwenn (Eds.), Vol. 2, Max Planck Gesellschaft, `http://solarphysics.livingreviews.org`

Gizon, L., Schunker, H., Baldner, C. S., Basu, S; Birch, A. C.; Bogart, R. S.; Braun, D. C.; Cameron, R.; Duvall, T. L.; Hanasoge, S. M.; Jackiewicz, J.; Roth, M.; Stahn, T.; Thompson, M. J. & Zharkov, S. (2009). Helioseismology of sunspots: A case study of NOAA Region 9787. *Space Science Reviews*, Vol. 144, 249–273

González Hernández, I.; Hill, F. & Lindsey, C. (2007). Calibration of the far sid seismic-holography signature of active regions. *Astrophys. J.*, Vol. 669, 1382–1389

Gonzĺez Hernández, I.; Hill, F.; Scherrer, P. H.; Lindsey, C. & Braun, D. C. (2009). On the success rate of the farside seismic imaging of active regions. *Space Weather* Vol. 8, S06002

Hudson, H.S.; Fisher, G. H. & Welsch, B. T. (2008). Flare energy and magnetic field variations. *Subsurface and Atmospheric Influences on Solar Activity, ASP Conf. Series* Vol. 383, R. Howe, R. Komm, K. S. Balasubramaniam & G. J. D. Petrie (Eds.), pp. 221–226, National Solar Observatory, Sunspot, NM, 16–20 April 2007, Astronomical Society of the Pacific

Kosovichev (2007). The cause of photospheric and helioseismic responses to solar flares: High-energy electrons or protons. *Astrophys. J. Letters*, Vol. 670, 65–68

Kosovichev, A. G.; Duvall, T. L. Jr. & Scherrer, P. H. (2000). Time-distance inversion methods and results. *Solar Phys.*, Vol 192, 159–176

Kosovichev, A. G. & Zharkova, V. V. (1998). X-ray flare sparks quake inside Sun. *Nature*, Vol 393, 317–318

Leighton, R. B.; Noyes, R.; Simon, G. (1962). Velocity fields in the solar atmosphere. Preliminary report. *Astrophys. J.*, Vol. 135, 474–499

Lindsey, C. & Braun, D. C. (1990). Helioseismic imaging of sunspots at their antipodes. *Solar Phys*, Vol. 135, 474–499

Lindsey, C. & Braun, D. C. (1997). Helioseismic Holography. *Astrophys. J.*, Vol. 485, 895–903

Lindsey, C. & Braun, D. C. (1999). Chromatic holography of the sunspot acoustic environment. *Astrophys. J.*, Vol. 510, 494–504

Lindsey, C. & Braun, D. C. (2000a). Seismic Images of the Far Side of the Sun. *Science*, Vol. 287, 1799–1800

Lindsey, C. & Braun, D. C. (2000b). Basic Principles of Solar Acoustic Holography. *Solar Phys.*, Vol. 192, 261–284

Lindsey, C. & Braun, D. C. (2005a). The acoustic showerglass. I. Seismic diagnostics of photospheric magnetic fields. *Astrophys. J.*, Vol. 620, 1107–1117

Lindsey, C. & Braun, D. C. (2005b). The acoustic showerglass. II. Imaging active-region subphotospheres. *Astrophys. J.*, Vol. 620, 1118–1131

Lindsey, C.; Cally, P. S. & Rempel, M. (2010). Seismic discrimination of thermal and magnetic anomalies in sunspot umbrae. *Astrophys. J.*, Vol. 719, 1144–1156

Lindsey, C. & Donea, A.-C. (2008). Mechanics of seismic emission from flares. *Solar Phys.*, Vol. 651, 227–639

Lindsey, C.; Braun, D. C.; Jefferies, S. M.; Woodard, M. F.; Fan, Y.; Gu, Y. & Redfield, S. (1996). Doppler acoustic diagnostics of subsurface solar magnetic structure. *Astrophys. J.*, Vol. 470, 636–646

Moradi, H., Baldner, C.; Birch, A. C.; Braun, D. C.; Cameron, R. H.; Duvall, T. L.; Gizon, L.; Haber, D.; Hanasoge, S. M.; Hindman, B. W.; Jackiewicz, J.; Khomenko, E.; Komm, R.; Rajaguru, P.; Rempel, M.; Roth, M.; Schlichenmaier, R.; Schunker, H.; Spruit, H. C.; Strassmeier, K. G.; Thompson, M. J. & Zharkov, S. (2010). Modeling the subsurface structure of sunspots. *Solar Phys.*, Vol. 267, 1–73

Moradi, H.; Donea, A.-C.; Lindsey, C.; Besliu-Ionescu, D. & Cally, P. S. (2007). Helioseismic analysis of the solar flare-induced sunquake of 2005 January 15. *Min. Roy. Astron. Soc.* Vol. 374, 1155–1163

Quemerais, E. & Bertaux, J.-L. (2006). Retrieving the solar wind mass flux latitude and cycle dependence with SWAM/SOHO data. *Proc. COSPAR Sci. Assem.*, Vol. 36, 1615, Beijing

Roddier, F. (1975). Principe de réalisation d'un hologramme acoustique de lo surface du Soleil. *Comptes Rendus, Serie B—Sciences Physiques*, Vol. 281, 93–95

Rhodes, E. J.; Ulrich, R. & Simon, G. W. (1977). Observations of nonradial p-mode oscillations on the Sun. *Astrophys. J.*, Vol. 218, 901–919

Rhodes, E. J.; Deubner, F.-L. & Ulrich, R. (1979). A new technique for measuring solar rotation. *Astrophys. J.*, Vol. 227, 629–637

Scherrer, P. H.; Bogart, R. S.; Bush, R. I.; Hoeksema, J. T.; Kosovichev, A. G.; Schou, J.; Rosenberg, W.; Springer, L.; Tarbell, T. D.; Title, A.; Wolfson, C. J.; Zayer, I. & the MDI Engineering Team (1995). The Solar Oscillations Investigation—Michelson Doppler Imager *Solar Phys.*, Vol. 162, 129–188y

Schunker, H.; Braun, D. C.; Lindsey, C. & Cally, P. S. (2008). Physical properties of wave motion in inclined magnetic fields within sunspot penumbrae. *Solar Phys.*, Vol. 251, 341–359

Skartlien, R. (2001). Imaging of acoustic wave sources inside the Sun. *Astrophys. J.*, Vol. 554, 488–495

Skartlien, R. (2002). Local Helioseismology as an inverse source-inverse scattering problem. *Astrophys. J.*, Vol. 565, 1348–1365

Spruit, H. C. & Bogdan, T. J. (1992). The conversion of p-modes to slow modes and the absorption of acoustic waves by sunspots. *Astrophys. J. Letters*, Vol. 391, 109–112

Stein, R. F.; Bogdan, T. J.; Carlsson, M.; Hansteen, V; McMurry, A; Rosenthal, C. S. & Nordlund, A . (2004). Theory and simulations of solar atmospheric dynamics. A joint view from SOHO and TRACE. Proc. SOHO13 Conference. *Waves, Oscillations and Small-scale Events in the Solar Atmosphere*, pp. 93–105, Palma de Mallorca, Balearic Islands (Spain), September–October 2003, ESA SP-547

Ulrich, R. (1970). The five-minute oscillations on the solar surface. *Astrophys. J.*, Vol. 162, 993–1002

Zernike, F. (1938). The concept of degree of coherence and its application to optical problems. *Physica*, Vol. 5, 785–795

5

Polarization Holographic Gratings Formed on Polymer Dispersed Liquid Crystals

Zharkova G. M., Petrov A. P., Streltsov S. A. and Khachaturyan V. M.
Khristianovich Institute of Theoretical and Applied Mechanics of the Siberian Branch of the Russian Academy of Science, Russia

1. Introduction

Development and investigation of new active optical media purposed for the reflection of optical information and also for the control of light propagation direction and properties in photonic devices remain essential under the conditions of rapid development of information technologies. Because of their optically active properties, polarization holographic gratings are promising in the production of high-function optical devices such as lights modulators, valves, polarization multiplexers and demultiplexers, polarized radiation splitters (Bunning et al., 2000; Cipparrone et al., 2000; Cipparrone et al., 2001a; Habraken et al., 1995; Ono et al., 2005; Wu et al., 2001; Yu et al., 2002). Such gratings are recorded on polarization-sensitive materials: photo-refractive crystals (Sturman et al., 1996), «guest-host» systems (Khoo, 1991; Slussarenko et al., 1997; Nikolova & Todorov, 1984), azopolymers (Naydenova et al., 1998; Labarthet & Rochon, 1999). The reports about the record of the polarization gratings in liquid-crystal composites (LC-composites) which do not contain azodyers, are comparatively recent (Cipparrone et al., 2001a; Mazzulla et al., 2004).

The LC-composites present a polymer matrix, within which the droplets of nematic liquid crystals (NLC) are dispersed. They are appealing by the fact that they combine the positive features of polymers (mechanical strength, stability, flexibility, low price) and unique physical peculiarities of liquid crystals (anisotropy of optical and dielectric properties). Such composite materials allow using a relatively simple one-stage technology of the optical record, a long-term retention of the polarization state, which provides the high polarization selectivity of the recorded polarization gratings. Moreover, the diffraction efficiency of the polarization holographic gratings recorded in the LC-composites can be controlled by electric and thermal fields.

The polarization holographic gratings form in a thin LC-composite layer as a result of the NLC-polymer phases separation during the photopolymerization caused by the superposition in the plane of the sample of two coherent waves polarized in the mutually perpendicular planes or two left- and right-circularly polarized waves. In this case, the light modulation is absent, and the changing phase difference between interfering waves results solely in the variation of the polarization state of the total light field. Since the NLC is sensitive to the total field direction and is able to orient in the light-field vector direction, a grating with repeating orientations of the NLC molecules in polymer droplets forms in the LC-composite after the exhibition.

The functionality of the formed gratings, effective control of their electric-optical properties depend on the NLC physical properties and are also related with the peculiarities of the supramolecular structure which is realized in the forming composite. Among the parameters governing the supramolecular structure, the following should be noted: the NLC and polymer type, LC – polymer ratio, technological process peculiarities (record energy, formation temperature).

Because of big amount of components in the pre-polymer system, it is necessary to study the balance between their concentrations and corresponding physical and chemical properties of the forming gratings, which dictate the grating controllability by the electric field. The influence of all mixture components on the properties of reflective and transmission gratings formed at the light photopolymerization with the modulated intensity has been described in the literature (Beev et al., 2006; Zharkova et al., 2007). In these cases, the periodical supramolecular structures form at competing processes of photopolymerization, diffusion, phase separation of the NLC and polymers and vitrification. When the polarization gratings form, the initial composition is illuminated by the light of constant intensity thus there are no concentration gradients of the components in dark and light bands of the interferential pattern. Therefore the processes of substance diffusion are absent. That is why the quality of the polarization gratings depends on the photopolymerization, NLC – polymer phase separation and vitrification. These processes significantly depend on the initial composition content and record conditions. Today, for the polarization gratings, there are no clear conditions of optimization of all processes involved in the formation of gratings with the maximum diffraction efficiency because of the lack of experimental data. The reason is that for each individual system there are its optimum component relations and exhibition conditions.

The peculiar structure of the NLC in such materials presumes the dependence of the polarization holographic gratings efficiency on the incident radiation polarization, as well as the difference in the polarization states of the diffracted and non-diffracted radiation, which enables to control the direction of the optical radiation propagation and polarization.

Since the anisotropic properties of the nematic liquid crystal, which is the component of the pre-polymer composition, are influenced by the temperature and external electric field, the peculiarities of thermal-optical and electric-optical switching of the polarization holographic gratings are of interest.

Thus, the purpose of this work is to study the influence of the formation conditions of the polarization holographic gratings, recorded in the LC-composites, on their diffraction efficiency to investigate the polarization selectivity of such gratings, as well as to study the peculiarities of the thermal-optical and electric-optical switching of the polarization holographic gratings and to demonstrate the possibility of using them as light modulators.

2. Theory

The polarization gratings record resulted from the superposition of two plane waves with mutually orthogonal polarizations and equal amplitudes E_0. The waves were polarized as follows: a) at the angles of $+45^0$ and -45^0 to the incidence plane; b) at the angles of 0^0 and 90^0 to the incidence plane (*s*- and *p*-polarization) (Fig.1).

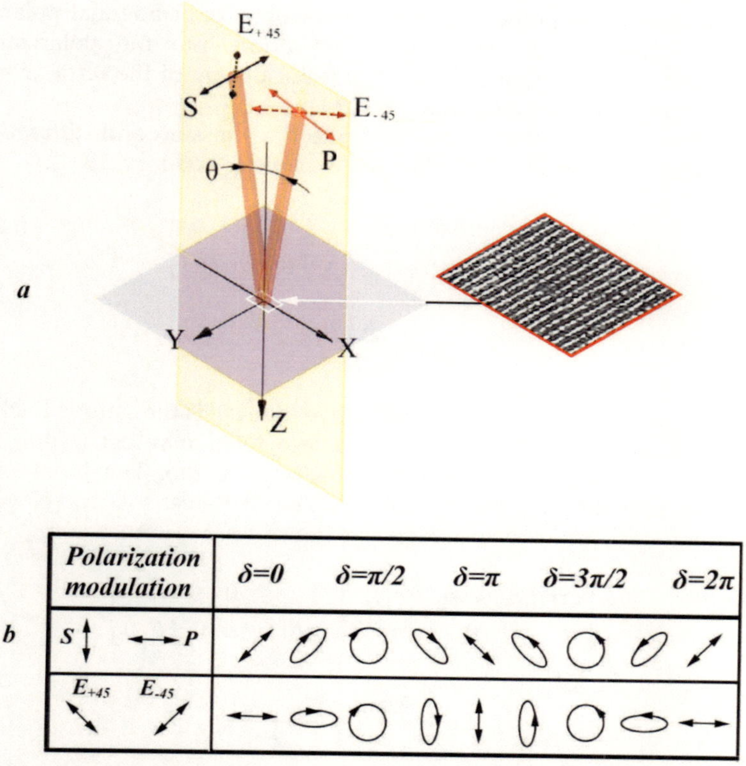

Fig. 1. Geometry of polarization holographic gratings record (a) and respective spatial polarization modulation along the axis *x* (b). The photo of the polarization holographic grating is obtained by the polarization microscopy method, polarizers are crossed.

The total field strength $E=E_1+E_2$ can be written with the Jones vector, for the case a):

$$E=\frac{2E_0}{\sqrt{2}}\cdot\begin{bmatrix}\cos\frac{\theta}{2}\cdot\cos\frac{\delta}{2}\\ i\cdot\sin\frac{\delta}{2}\end{bmatrix}; \tag{1}$$

where

$$\delta=\frac{4\pi\cdot x\cdot sin(\frac{\theta}{2})}{\lambda}; \tag{2}$$

is the difference of interfering waves phases; λ is the wave length; θ is the angle of interfering waves convergence;

for the case b):

$$\mathbf{E}=E_0\cdot\begin{bmatrix}cos\frac{\delta}{2}\\ isin\frac{\delta}{2}\end{bmatrix}; \tag{3}$$

Let us presume that at the superposition of two waves with orthogonal polarizations, the varying refractive index gives the major contribution into the polarization grating formation. Moreover, presume that only the linear anisotropy of the refractive index takes place for the used composition at the grating record.

In this case, the Jones matrix, which describes the transmission and diffraction from the recorded grating, can be determined as follows (Nikolova & Todorov, 1984):

$$\mathbf{T_1} = \begin{bmatrix} exp\{(i\Delta\phi)cos\delta\} & 0 \\ 0 & exp\{-(i\Delta\phi)cos\delta\} \end{bmatrix}; \tag{4}$$

where

$$\Delta\phi = \frac{\pi\Delta nd}{\lambda}; \tag{5}$$

Δn is the photo-induced modulation of the refractive index; d is the sample thickness;

While the polarization grating is forming, a surface relief may occur during the phase separation process. The incident radiation diffracts on it, too. The Jones matrix, which describes the diffraction of such a surface relief, can be written as follows (Naydenova et al., 1998):

$$T_2 = \begin{bmatrix} exp\{(i\Delta\psi)cos(\delta+\delta_1)\} & 0 \\ 0 & exp\{(i\Delta\psi)cos(\delta+\delta_1)\} \end{bmatrix}; \tag{6}$$

where

$$\Delta\psi = \frac{2\pi \cdot d_1}{\lambda} \cdot \frac{(n_p - n_a)}{2}; \tag{7}$$

is the value of the phase modulation caused by the surface relief; d_1 is the surface modulation depth; n_p is the average refractive index of the LC-composite; n_a is air refractive index; δ_1 is the phase shift between the relief and polarization gratings.

The field of the diffracted wave can be defined as follows (Cipparrone et al., 2001b):

$$E_{out} = T_2 T_1 E_{in}; \tag{8}$$

where

$$E_{in} = E_{0i} \begin{bmatrix} cos\alpha \\ sin\alpha \end{bmatrix}; \tag{9}$$

is the Jones vector describing the incident wave; E_{0i} is the amplitude of the incident wave; a is the angle between the direction of the test beam polarization and wave vector of the grating.

The polarization properties of the beam diffracted into the m–order are determined by the Fourier components $E_{diffm}(a)$ on the paraxial approximation (Tervo & Turunen, 2000):

$$E_{diffm}(a) = \frac{1}{\Lambda} \times \int_0^{\Lambda} E_{out}(x) exp(-i2\pi mx / \Lambda)dx; \tag{10}$$

where

$$\Lambda = \frac{\lambda}{2sin(\theta/2)} \tag{11}$$

is the grating period;

the diffraction efficiency of the polarization grating can be defined as:

$$\eta_m = exp(-\sigma \times d) \times \frac{\left|E_{diffm}(a)\right|^2}{\left|E_{in}\right|^2}; \tag{12}$$

where σ is the extinction ratio.

3. Formation of the polarization gratings and experimental methods

Polarization gratings were recorded as a result of superposition of two plane waves of similar intensity with mutually orthogonal linear polarizations. In one case, the waves were polarized vertically and horizontally (**s** and p), in the other case – at the angle of ± 45^0 ($\mathbf{E}_{+45}$ and $\mathbf{E}_{-45}$) to the incident plane (Fig. 1). At the superposition of the waves with mutually orthogonal polarizations, solely the polarization state varies in the sample plane (XZ), and the modulation intensity is absent.

Fig. 2 shows the schematic of the polarization holographic grating record. A semi-conductor laser with λ = 658 nm and power of 80 mW was used as a radiation source. The radiation power density in the sample plane was 42 mW/cm^2. The angle of convergence of the interfering light beams was $\theta = 1.3^0 - 2^0$. The period of the obtained gratings was equal to 19 - 30 μm, which corresponded to the spatial frequency of 53 - 34 mm^{-1}.

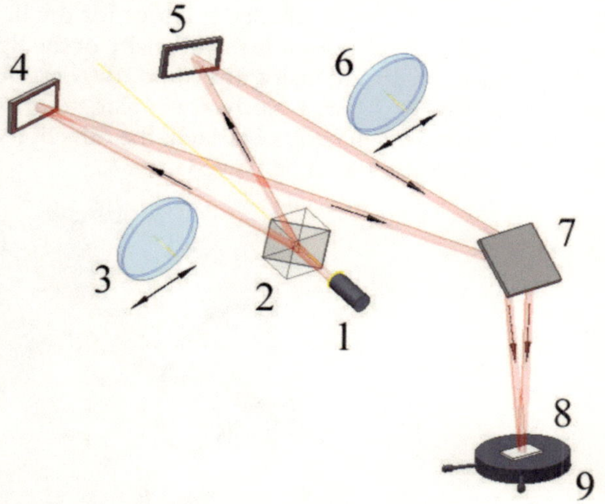

Fig. 2. Schematic of the facility for polarization holographic gratings record. 1 – laser diode ML101GR (λ = 658 nm); 2 – Wollaston prism; 3, 6 – plates λ/2; 4, 5, 7 – mirrors; 8 – sample; 9 – thermostatic table.

The initial prepolymer composition presented a mixture of an acrylate monomer, NLC, cross-linking component and the photo-initiating system which included a dye with the maximum absorption in the visible spectrum, and a co-initiator. Pentaerythritol tetraacrylate (Sigma-Aldrich) (n_p=1,487) was used as the monomer, and the mixture BL087 (Merck) (n_o=1.527, n_e=1.797, $\Delta\varepsilon$>0, T_{NI} =100 ^{o}C) was applied as the NLC. In the composition, N-vinylpyrrolidone served as the cross-linking component and a solvent for the dye and co-

initiator. Methylene-blue was used as the dye; it guaranteed the maximum light absorption within the wave length range of 650-680 nm. The NLC concentration in the prepolymer composition was 25-55 mass %; the dye concentration was 0.3 mass %. The composition was prepared as a homogeneous solution which was deposited between two glass slides coated with a transparent conductor. Sample thickness was 10 - 50 µm.
At the radiation diffraction on the formed polarization gratings, we observed the high-order diffraction (Fig. 3). The Klein's parameter for the such gratings is Q:

$$Q = 2\pi\lambda d / (n\Lambda^2) \leq 1; \tag{13}$$

(Raman-Nath regime), i.e. we deal with plane holograms.

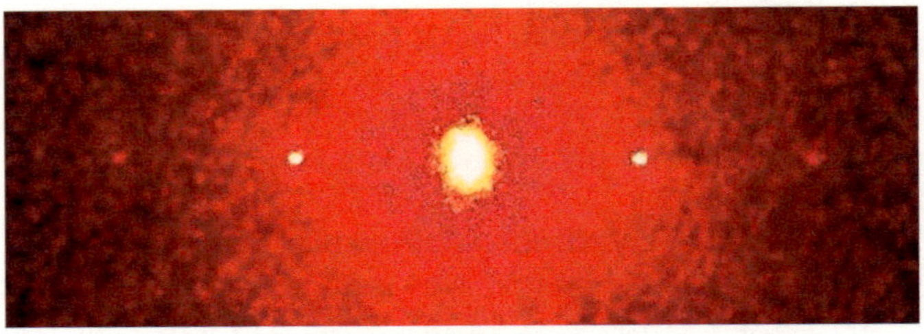

Fig. 3. Diffraction pattern observed at the diffraction of the polarization grating.

The quality of the recorded polarization holographic gratings was evaluated by their diffraction efficiency. The diffraction efficiency of such gratings for the light diffracted into the ±1 order, was defined as $\eta_{\pm 1} = I_{\pm 1}/I_0$, where $I_{\pm 1}$ is the intensity of the light diffracted into the first order, I_0 is the incident light intensity the onto the grating. Fig. 4 shows schematically the experimental facility for the study the polarization selectivity of the polarization grating. The probe beam was linearly polarized, the direction of its polarization formed the angle α with the direction of the grating wave vector **k**.

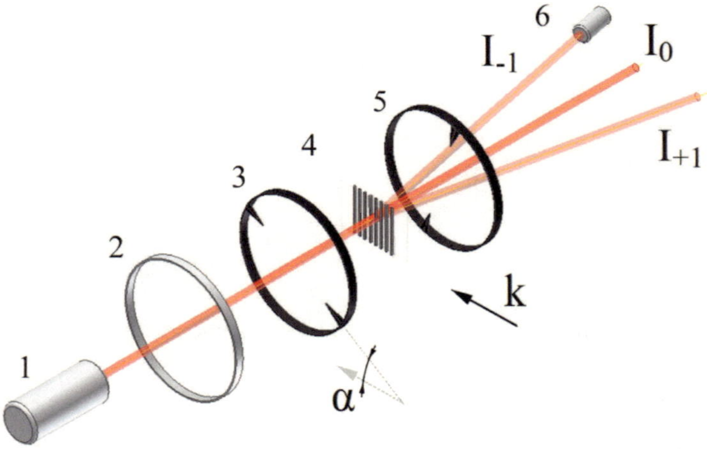

Fig. 4. Schematic of the facility to study the polarization selectivity of the polarization gratings. 1 – semiconductor laser (λ = 658 nm); 2 - λ/4 plate; 3 – polarizer; 4 – sample under study; 5 – analyzer; 6 – photodetector.

To study the thermal-optical switching, the studied sample was set into a thermostatic cell.

4. Experimental results

It is possible, varying the content of the prepolymer composition and its lighting mode, to control the kinetics of photopolymerization and components phase separation, and hence to influence the supramolecular structure of the obtained composite and, consequently, the properties of the forming polarization holographic grating. Two factors have the most significant effect on the formed composite morphology: a) the NLC – polymer ratio in the initial prepolymer composition; b) exposure energy.

The study of the effect of the NLC concentration in the initial prepolymer composition has revealed that the gratings with better efficiency form within quite a narrow NLC concentration interval (35 - 38%), and any deviation from it results in the noticeable deterioration of their quality (Fig. 5).

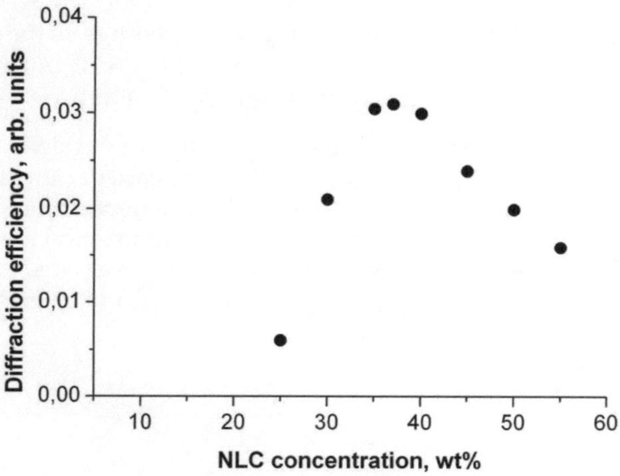

Fig. 5. Diffraction efficiency of polarization holographic gratings vs the NCL concentration in the initial prepolymer composition. Sample thickness is 20 μm. Spatial frequency of the grating is 34 mm^{-1}.

The studied polarization gratings result from the polarization and phase separation processes. At low NLC concentrations (~ 25%), the process of the phase separation of polymer and liquid-crystal phases is at the beginning, and at the concentrations of 45 – 55% it is almost over, which causes strong light dispersion on numerous polymer-NLC interfaces and hence results in increasing dispersion noise and decreasing diffraction efficiency of the grating.

Fig. 6 presents the dependence of the diffraction efficiency of the polarization holographic grating, formed in the LC-composite on the record energy density for the grating with the spatial frequency of 34 mm^{-1}. It is evident that the diffractive efficiency rises along with the record energy up to ~3%, then the growth decelerates and saturation is observed at η = 3.5% (for the sample of 20 μm). The threshold record energy corresponding to the diffraction efficiency of 1% is 0.65 J/cm^2, and the total record energy at which the maximal diffraction efficiency is reached is equal to 2.5 J/cm^2. The sample thickness influences these values. The smaller the thickness of the LC-composite, the lower energy density is needed to finish the orientation processes of the NLC and processes of the phase separation of polymer and liquid-crystal phases at the specific radiation power density.

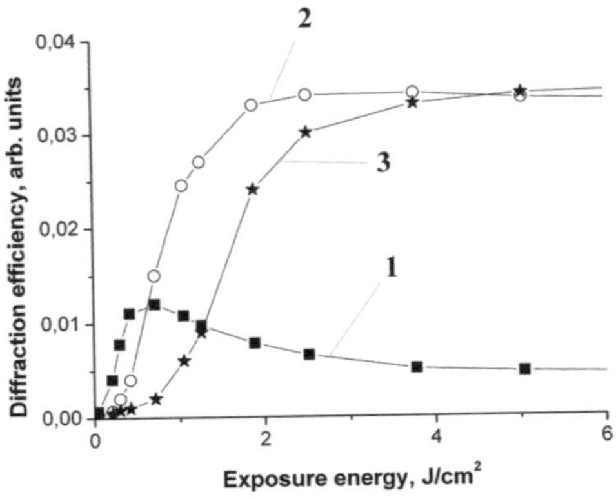

Fig. 6. Diffraction efficiency of the polarization holographic gratings vs the energy density for LC-composites of various thickness. 1 – thickness of 10 μm; 2 – 20 μm; 3 – 50 μm. NLC concentration is 37 mass %. Spatial frequency of the grating is 34 mm^{-1}.

As the spatial frequency of the recorded polarization grating grows (along with the growing angle of interfering beams convergence), we observe the increase of the total energy needed for its forming, regardless the grating thickness (Fig. 7). The dependence of the diffraction effectiveness on the record energy density features the maximum, which position and value depend on the grating spatial frequency. Moreover, in our experiments, for the chosen pre-polymer composition, the polarization gratings with the spatial frequency above 60 mm-1 did not form even at significant record energies.

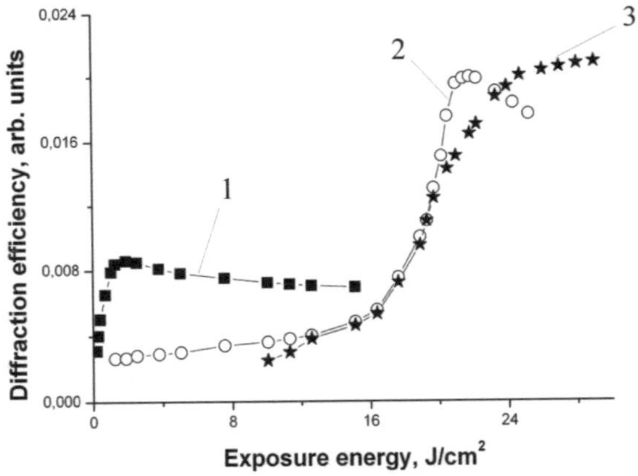

Fig. 7. Diffraction efficiency of the polarization holographic gratings vs the energy density for LC-composites of various thickness. 1 – thickness of 10 μm; 2 – 20 μm; 3 – 50 μm. NLC concentration is 37 mass %. Spatial frequency of the grating is 53 mm^{-1}.

At the polarization grating formation, the process of phase separation is temperature-sensitive; along with the temperature growth, the NLC and monomer viscosity decrease, the phase separation of the NLC and polymer is faster and better, and the polymerization velocity increases. Moreover, the temperature influences the process of NLC orientation in the droplets. Fig. 8 shows the dependence of the diffraction efficiency of the polarization holographic gratings on the formation temperature for the temperature range: from the room temperature to the point of NLC transition in an isotropic liquid.

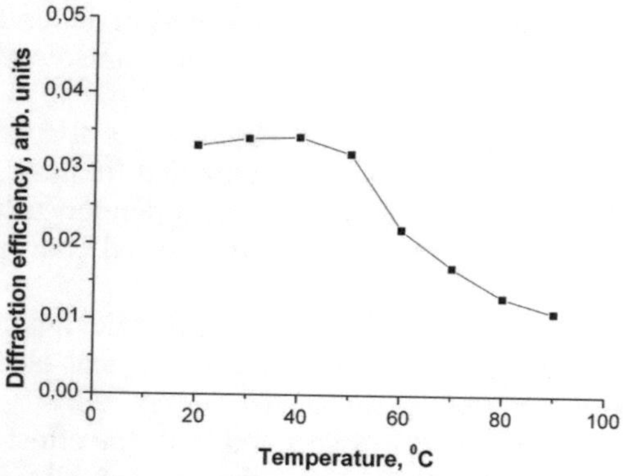

Fig. 8. Diffraction efficiency of the polarization holographic gratings vs the formation temperature. Sample thickness is 20 μm. Spatial frequency of the grating is 34 mm^{-1}.

Evident that the maximum diffraction efficiency is reached at the formation temperature of 35 - 45 ^{0}C. This is apparently the temperature range in which the processes of polymerization, NLC-polymer phase separation and NLC orientation proceed in balance under the action of the total field. At further temperature growth, these processes begin to compete with each other, and above 50 ^{0}C, the diffraction efficiency decrease significantly.
The sensitivity of the material for hologram records is commonly characterized by the parameter (Ortler et al., 1989):

$$g = \eta^{1/2} / W, \tag{14}$$

where W is the record energy density; η is the corresponding diffraction efficiency.
For our samples, this parameter lies within the limits of 0.04 – 0.15 cm^2/J, regarding the thickness, which is lower than in photo-reactive materials (0.2 – 0.5 cm^2/J) (Ducharme et al., 1991; Eich & Wendorff, 1990).
At the electric field imposing, the NLC molecules are oriented along the field (since $\Delta\varepsilon > 0$), and for the radiation incident on the polarization holographic grating, the refractive index modulation is very small, and hence there is no radiation diffraction on such a grating.
Fig. 9 presents the optical response of the radiation diffracted on the polarization holographic grating, to the directing alternating electric field. It is evident that the condition of $n = -n$, satisfied for the director in the NLC, provides doubling frequency of the directing field by the optical response.

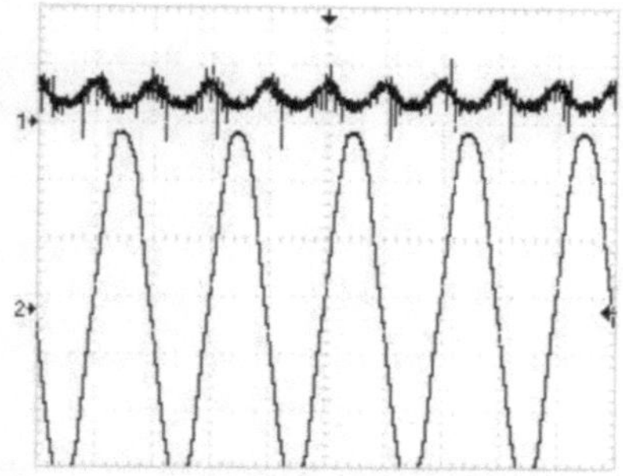

Fig. 9. Optical response of the polarization holographic grating (1) to the alternating voltage with frequency of 50 Hz (2).

The value of the electric field needed for the LC re-orientation depends on many factors: LC-composite thickness, droplets size, optical and dielectrical anisotropy (Zharkova & Sonin, 1994).
Fig. 10 shows the dependence of the diffraction efficiency of the polarization holographic grating on the applied alternating electric voltage with the frequency of 50 Hz for various conditions of the grating formation. Evident that the dependence has a threshold nature, and the curve inclination is caused by the polidispersity of NLC droplets formed in the LC-composite.
Moreover it is seen that for the LC-composites of the same thickness, the mode of formation of the polarization holographic gratings influences the critical electric fields. The critical field is smaller for the gratings formed at lower energy density than for the gratings recorded at the higher energy density. It is associated with the effect of the radiation energy on the LC-composite morphology and in particular on the NLC droplets size, and the critical field is inversely proportional to the droplet size (Zharkova & Sonin, 1994).

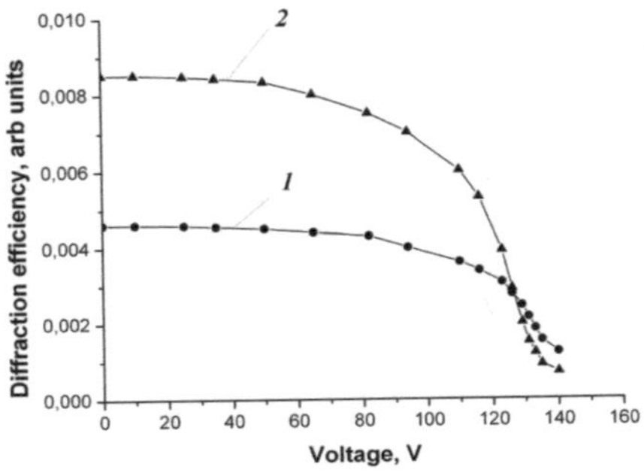

Fig. 10. Diffraction efficiency of the polarization holographic gratings vs the applied voltage. 1 – record energy density 1.2 J/cm^2; 2 – record energy density 0.45 J/cm^2. LC-composite thickness is 10 μm. Spatial frequency of the grating is 53 mm^{-1}.

As follows from Figure 11, the critical field value highly depends on the LC-composite thickness, in which the polarization holographic grating forms. The critical electric field rises along with the thickness.

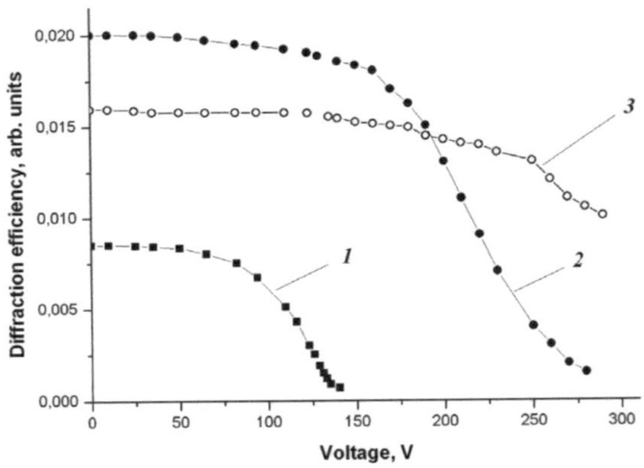

Fig. 11. Diffraction efficiency of the polarization holographic gratings vs the applied voltage for various thicknesses of the LC-composites. 1 – thickness of 10 μm; 2 – 20 μm; 3 – 50 μm. Spatial frequency of the grating is 53 mm^{-1}.

It is also possible to control the diffracted radiation by means of the temperatures, since the physical properties of the NLC are sensitive to this parameter. In this context, we can observe the thermooptical switching into the state when the grating is "erased". Fig. 12 presents the dependence of the diffraction efficiency on the temperature.

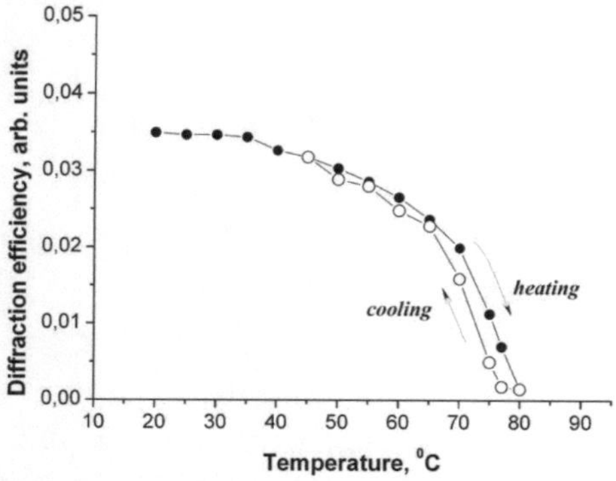

Fig. 12. Diffraction efficiency of the polarization holographic gratings vs the temperature. Sample thickness is 20 μm. Spatial frequency of the grating is 34 mm-1.

As is seen from the Figure, as the temperature grows, the diffraction efficiency decreases and at ~80 ^{0}C the diffraction properties of the grating disappear, i.e. the nematic liquid crystal transforms into the isotropic liquid. The temperature of the NLC – isotropic liquid transformation is considerably lower for the samples under study, than it is for the pure NLC used in the initial composition, which may result from the NLC pollution by the substances of the prepolymer composition, above all, by the polymer. At the grating cooling, as is seen from the Figure, its diffraction properties are recovered with a slight hysteresis.

Special structure of the NLC occurring at the polarization holographic gratings record causes the dependence of the diffraction efficiency on the incident radiation polarization and also the difference between the states of the diffracted and non-diffracted radiation.

In both cases, as the polarization grating is recorded by the *s*- and *p*-polarized beams, or it is recorded by the beams polarized at the angles of ± 45^0, the intensities of the beams diffracted in the first order I_{+1} and I_{-1} were similar. Fig. 13 shows the dependence of the diffraction efficiency of the polarization holographic grating on the azimuth polarization angle of the probe beam α.

It is seen that the recorded gratings have the polarization selectivity, the diffraction efficiency highly depends on the probe beam polarization direction. At the incident *s* - or *p* - polarized probe beam on the grating, which is formed also by the *s* - and *p* – polarized beams, the non-diffracted beam (the 0 order) had the same polarization as the incident one, and the beams diffracted in the first order had the orthogonal polarization, i.e. if the incident beam had the *s* – polarization, the beams diffracted in the first order had the *p*-polarization and vice versa. If the sample was beamed by the radiation polarized at the angle of ±45^0, the non-diffracted beam (the 0 order), as well as the beams diffracted in the first order, had the same polarization which coincided with the polarization of the light incident on the sample.

For the gratings formed by the beams polarized at the angles of ±45^0, with the probe incident beam polarized at the angle of +45^0 or -45^0, the beams diffracted in the first order had the orthogonal polarization, i.e. -45^0 and +45^0, respectively. The non-diffracted beam (the 0 order) had the same polarization as the incident one. If an *s* - or *p* –polarized beam beamed such a grating, the non-diffracted beam (the 0 order), as well as the beams diffracted in the first order, had the same polarization coinciding with the polarization of the light incident on the sample.

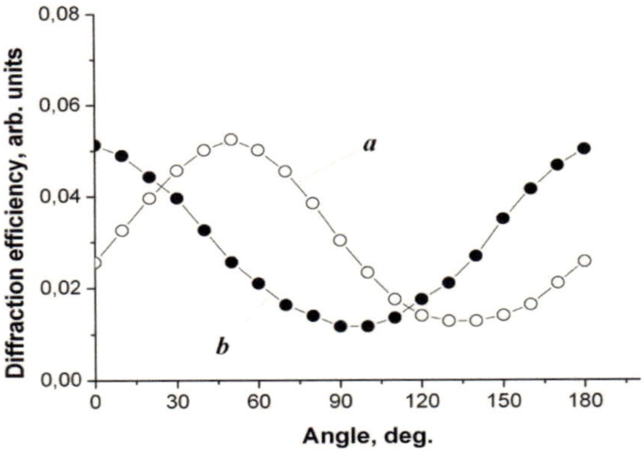

Fig. 13. Diffraction efficiency of the polarization holographic grating vs the azimuth angle of the probe beam polarization: a – the grating is recorded by the s- and p-polarized beams; b – the grating is recorded by the beams polarized at the angle of ± 45^0 to the incident plane. Spatial frequency of the gratings is 34 mm^{-1}. Solid curves are the calculations of the diffraction efficiency by using Eq. (12). Sample thickness is 20 μm

The properties of the polarization gratings described above enable to control the diffracted radiation. Fig. 14 presents the photos of the diffraction patterns obtained at the diffraction of the s – polarized probe beam on the grating recorded by the s - and p – polarized beams. When the polarizer and analyzer are crossed (Fig. 3), the non-diffracted beam (the 0 order) can be suppressed, and when they are parallel, the same is valid for the beams diffracted in the first order.

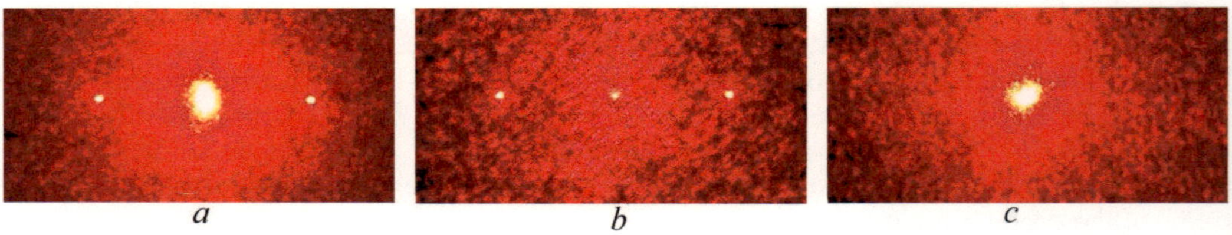

Fig. 14. Photos of diffraction patterns. *a* – without analyzer; *b* – crossed polarizer and analyzer; *c* – parallel polarizer and analyzer.

5. Conclusions

Thus, the one-stage process of record involving the polarization holography method results in the high-structured polarization holographic gratings with controllable supramolecular structure in the LC-composites. The diffraction efficiency of such gratings depends on the grating record energy, its formation temperature, NLC concentration in the prepolymer composition. It has been established experimentally that the gratings with the highest diffraction efficiency form at the temperatures of 30 – 45 ^{0}C, the NLC concentration in the prepolymer composition should be 35 – 38 mass %. The threshold record energy corresponding to the diffraction efficiency of 1 % is 0.65 J/cm^2, and the total record energy at which the maximal diffraction efficiency is reached is 2.5 J/cm^2 (for the grating up to 20 μm, with the spatial frequency of 34 mm^{-1}). The total energy needed for the grating formation increases along with the spatial frequency of the recorded polarization grating, regardless its thickness.

The optical properties of the polarization holographic gratings can be controlled by the electric field; the doubling of the control voltage frequency by the optical response is guaranteed. The diffraction efficiency of such gratings and critical electric fields depends on the LC-composite thickness and record energy density of the grating. Increase of both parameters results in the critical field increase.

As the grating is heated up to ~80 ^{0}C, its diffraction properties disappear. The experiment has revealed the reduction of the NLC – isotropic liquid transformation temperature as compared to the pure NLC case, which may result from the NLC pollution by the substances of the prepolymer composition, above all, by the polymer. At the grating cooling, as is seen from the Figure, its diffraction properties are recovered with a slight hysteresis.

The diffraction efficiency of the polarized holographic gratings depends on the azimuth angle of the probe beam polarization. It has been established experimentally that the highest diffraction efficiency corresponds to the angle $\alpha = 45^0$ for the gratings formed by the s- and p-polarized beams and the angle of 0^0 in the case of the beams polarized at the angles of $\pm 45^0$. The possibility of "suppressing" of the beams diffracted on the polarization grating has been demonstrated.

6. References

Beev K., Criante L., Lucchetta D., Simoni F., Sainov S. Total internal reflection holographic gratings recorded in polymer-dispersed liquid crystals. *Opt. Comm.* 2006. Vol. 260, pp. 192-195.

Bunning T.J., Natarajan L.V., Tondiglia V.P., Sutherland R.L. *Annu. Rev. Mater. Sci.* – Vol. 30, p. 80 (2000).

Cipparrone G., Mazzulla A., S.P. Palto, S.G. Yudin, L.M. Blinov Permanent polarization gratings in photosensitive Langmuir-Blodgett films. *Appl. Phys. Lett.* – Vol. 77, №14 (2000), pp. 2106-2108.

Cipparrone G., Mazzulla A., Russo G. Diffraction gratings in polymer-dispersed liquid crystals recorded by means of polarization holographic technique. *Appl. Phys. Lett.* – Vol. 78, №9 (2001), pp. 1186-1188.

Cipparrone G., Mazzulla A., Russo G. Diffraction from holographic gratings in polymer-dispersed liquid crystals recorded by means of polarization light patterns. *J. Opt. Soc. Am. B.* 2001. Vol. 18, pp.1821-1826.

Ducharme S., Scott J.C., Twieg R.J., Moerner W.E. Observation of the photorefractive effect in a polymer. *Phys. Rev. Lett.* 1991. Vol. 66, pp. 1846-1849.

Eich M., Wendorff J.H. Laser-induced gratings and spectroscopy in monodomains of liquid-crystalline polymers. *J. Opt. Soc. Am. B* 1990. Vol. 7, pp. 1428-1436.

Habraken S., Michaux O., Renotte Y., Lion Y. Polarizing holographic beam splitter on a photoresist. *Opt. Lett.* Vol. 20, (1995), pp. 2348-2350.

Khoo I.C. Temporal dependence of optical nonlinearities of nematic liquid crystals and the unusual polarization dependence of self diffraction in 4-4'bis(heptyloxy) azoxybenzene. *Mol. Cryst. Liq. Cryst.* – Vol. 207 (1991), pp. 317-329.

Labarthet F. Lagugne, Rochon P. Polarization analysis of diffracted orders from a birefringence grating recorded on azobenzene containing polymer. *Appl. Phys. Lett.* – Vol. 75 (1999), pp. 1377-1379.

Mazzulla A., Pagliusi P., Provenzano C., Russo G., Carbone G., Cipparrone G. Surface relief ratings on polymer dispersed liquid crystals by polarization holography. *Appl. Phys. Lett.* – Vol. 85, №13 (2004), pp. 2505-2507.

Naydenova I., Nikolova L., Todorov T., Holme N.C.R., Ramanujam P.S., Hvilsted S. Diffraction from polarization holographic recording with surface relief in side-chain azobenzene polyesters. *J. Opt. Soc. Am. B*. 1998. Vol. 15, pp. 1257-1265.

Nikolova L., Todorov T. Diffraction efficiency and selectivity of polarization holographic Recording. *Opt. Acta.* – Vol. 31 (1984), pp. 579-588.

Ono H., Takahashi F., Emoto A., Kawatsuki N. Polarization holograms in azo dye-doped polymer dissolved liquid crystal composites. *J. of Appl. Phys.* – Vol. 97, 053508 (2005).

Ortler R., Brauchle C., Miller A., Riepl G., Reversible holographic-optical data storage in cholersteric liquid-crystalline siloxanes. Makromol. Chem. Rapid Commun. 1989. Vol. 10, pp. 189.

Slussarenko S., Francescangeli O., Simoni F. High resolution polarization gratings in liquid crystals. *Appl. Phys. Lett.* – Vol. 71 (1997), pp. 3613-3615.

Sturman B.I., Goul'kov M.Yu., Odoulov S.G. Phenomenological analysis of the paramertric scattering processes in photorefractive crystals. *J. Opt. Soc. Am. B*, - Vol. 13 (1996), pp. 577-583.

Tervo J., Turunen J. Paraxial-domain diffractive elements with 100% efficiency based on polarization gratings. *Opt. Lett.*, Vol. 25, pp. 785-786 (2000).

Wu Y.L., Natansohn A, Rochon P. Photoinduced birefringence and surface relief gratings in novel polyurethanes with azobenzene groups in the main chain. *Macromolecules*. – Vol. 34, pp. 7822-7828 (2001).

Yu W., Konishi T., Hamomoto T., Toyota H., Yotsuya T., Ichioka Y. Polarization-multiplexed diffractive optical elements fabricated by subwavelength structures. *Appl. Opt.*, Vol. 41, №1 (2002), pp. 96-100.

Zharkova G. M., Petrov A.P., Samsonova I.V., Streltzov S.A., Khachaturjan V.M. The infuence of the supramolecular structure of liquid-crystal composites on their optical properties. *Liquid crystals and their practical application.* 2007. Issue 3. pp. 45-58.

Zharkova G.M., Sonin A.S. Liquid-crystal composites. / Novosibirsk: Nauka, 1994, pp. 214.

6

Numerical Methods for Near-Field Acoustic Holography over Arbitrarily Shaped Surfaces

Nicolas P. Valdivia
Naval Research Laboratory
Washington, DC
United States of America

1. Introduction

This chapter is devoted to the numerical methods of the technique near-field acoustic holography (NAH) introduced by Williams & Maynard (1980). As explained in the book of Williams (1999), NAH has had a tremendous success analyzing source regions with geometries which conform closely to one of the separable geometries of the acoustic wave equation; for example, planar, cylindrical, and spherical geometries. In these analyses the pressure field radiated from an object is measured on an imaginary surface outside (for exterior problems) or inside (for interior problems) the source region. This measured pressure is transformed into a hologram of different frequencies that are used to reconstruct the acoustic field on the body of the source. The solution of this inverse problem in these geometries relies on the expansion of the pressure field in terms of a complete set of eigenfunctions along with a knowledge of the analytical form of the Neumann or Dirichlet Green's function. The reconstructions are very efficient, requiring only fractions of a second of computation time per frequency.

Source regions with boundaries which vary appreciably in shape from one of these separable geometries require numerical methods that will be based on boundary integral representations of the solution as in Colton & Kress (1983; 1992). This chapter discusses the available results in the acoustic literature for the integral representations and the numerical discretization methods, and finally the regularization methods used for this inverse problem.

2. Integral representations

Let G be a domain in $\mathbb{R}^3$, interior to the boundary surface Γ (as shown in Fig. 1) where we assume that Γ is allowed to have edges and corners. Similarly we will denote as G^+ the region outside of G that shares the same boundary Γ. For a time-harmonic ($e^{-i\omega t}$) disturbance of radial frequency ω the sound pressure p satisfies the homogeneous Helmholtz equation in G (or G^+)

$$\Delta p + k^2 p = 0, \qquad (1)$$

where p is given in N/m^2 units, $k = \omega/c$ is the wave number and c the constant for the speed of sound given in m/s units. Here $\omega = 2\pi f$, f is the frequency given in Hz. A solution p that

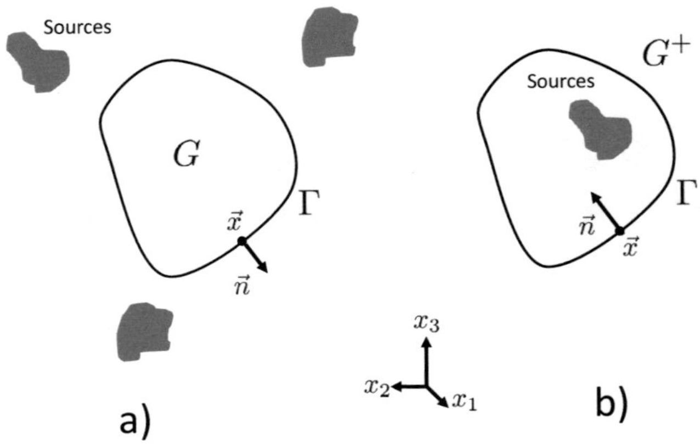

Fig. 1. Setup for the acoustic problem, a)interior problem and b)exterior problem.

satisfies Equation 1 in G^+ satisfies the Sommerfeld radiation condition

$$\lim_{r\to\infty} r\left\{\frac{\partial p}{\partial r} - ikp\right\} = 0, \quad r = |\vec{x}|, \tag{2}$$

where $\vec{x} = (x_1, x_2, x_3)$ and $|\vec{x}| = \sqrt{x_1^2 + x_2^2 + x_3^2}$.
At the boundary Γ the normal vibration v is related to the acoustic pressure p by Euler's equation

$$i\rho\omega v = \frac{\partial p}{\partial \vec{n}}, \tag{3}$$

where ρ is the mean fluid density given in kg/m^3 units and $\vec{n}$ is the vector unit normal with direction shown in Fig. 1.

2.1 Integral operators

We define the integral operators as in Colton & Kress (1983; 1992)

$$(S_\Gamma\varphi)(\vec{x}) := \int_\Gamma \Phi(\vec{x},\vec{y})\varphi(\vec{y})dS(\vec{y}), \quad \vec{x} \in \mathbb{R}^3, \tag{4}$$

$$(D_\Gamma\varphi)(\vec{x}) := \int_\Gamma \frac{\partial\Phi(\vec{x},\vec{y})}{\partial\vec{n}(\vec{y})}\varphi(\vec{y})dS(\vec{y}), \quad \vec{x} \in \mathbb{R}^3\backslash\Gamma, \tag{5}$$

$$(K_\Gamma\varphi)(\vec{x}) := \left(\frac{\partial S_\Gamma}{\partial\vec{n}}\varphi\right)(\vec{x}) := \int_\Gamma \frac{\partial\Phi(\vec{x},\vec{y})}{\partial\vec{n}(\vec{x})}\varphi(\vec{y})dS(\vec{y}), \quad \vec{x} \in \mathbb{R}^3\backslash\Gamma, \tag{6}$$

$$(H_\Gamma\varphi)(\vec{x}) := \left(\frac{\partial D_\Gamma}{\partial\vec{n}}\varphi\right)(\vec{x}) := \int_\Gamma \frac{\partial^2\Phi(\vec{x},\vec{y})}{\partial\vec{n}(\vec{x})\partial\vec{n}(\vec{y})}\varphi(\vec{y})dS(\vec{y}), \quad \vec{x} \in \mathbb{R}^3\backslash\Gamma, \tag{7}$$

where Φ is the *free space Green's function* to the Helmholtz equation

$$\Phi(\vec{x},\vec{y}) = \frac{\exp(ik|\vec{x}-\vec{y}|)}{4\pi|\vec{x}-\vec{y}|}. \tag{8}$$

In this work we will not discuss the smoothness properties of the operators. A detailed description of these properties is given in Colton & Kress (1983) and McLean (2000). The most important property for the purpose of NAH will be the evaluation of an operator S_Γ at $\vec{x} \in \Gamma$. We describe this as

$$(S_\Gamma \varphi)^{\pm}(\vec{x}) := \lim_{h \to 0} (S_\Gamma \varphi)(\vec{x} - h\vec{n}(\vec{x})),$$

where we will understand as the uppercase "+" the exterior problem with the direction of $\vec{n}$ as in Fig. 1b, while uppercase "−" the interior problem with the direction of $\vec{n}$ as in Fig. 1a. We keep this notation for all operators in Equations 4-7. Having explained the notation we have the *continuity relations*

$$(S_\Gamma \varphi)^{+}(\vec{x}) := (S_\Gamma \varphi)^{-}(\vec{x}), \tag{9}$$

$$(H_\Gamma \varphi)^{+}(\vec{x}) := (H_\Gamma \varphi)^{-}(\vec{x}), \tag{10}$$

and the *jump relations*

$$(D_\Gamma \varphi)^{\pm}(\vec{x}) := \int_\Gamma \frac{\partial \Phi(\vec{x}, \vec{y})}{\partial \vec{n}(\vec{y})} \varphi(\vec{y})\, dS(\vec{y}) \pm \Omega(\vec{x})\varphi(\vec{x}), \tag{11}$$

$$(K_\Gamma \varphi)^{\pm}(\vec{x}) := \int_\Gamma \frac{\partial \Phi(\vec{x}, \vec{y})}{\partial \vec{n}(\vec{x})} \varphi(\vec{y})\, dS(\vec{y}) \mp \Omega(\vec{x})\varphi(\vec{x}), \tag{12}$$

where $\Omega(\vec{x})$ is the solid angle coefficient given by the integral formula

$$\Omega(\vec{x}) = -\int_\Gamma \frac{\partial}{\partial \vec{n}(\vec{y})} \left(\frac{1}{4\pi |\vec{x} - \vec{y}|} \right) dS(\vec{y}), \quad \vec{x} \in \Gamma. \tag{13}$$

2.2 Integral formulations

The classical Helmholtz-Kirchhoff integral equation described in Colton & Kress (1983) is known as the integral representation of the pressure by the *direct formulation*. This classical representation has been used for purposes of solving NAH by several authors like Kim & Ih (1996; 2000); Kim & Lee (1990); Maynard (1988); Nelson & Yoon (2000); Sureshkumar & Raveendra (2001); Veronesi & Maynard (1989); Williams (2001); Yoon & Nelson (2000). The sound pressure in G^+ by the direct formulation has the integral representation

$$p(\vec{x}) = \begin{cases} i\rho\omega\,(S_\Gamma v)(\vec{x}) - (D_\Gamma p)(\vec{x}), & \vec{x} \in G^+, \\ 0, & \vec{x} \in G, \end{cases} \tag{14}$$

and in G

$$p(\vec{x}) = \begin{cases} 0, & \vec{x} \in G^+, \\ (D_\Gamma p)(\vec{x}) - i\rho\omega\,(S_\Gamma v)(\vec{x}), & \vec{x} \in G. \end{cases} \tag{15}$$

The *indirect formulations* are based on representations to the solution of Equation 1 like the single source or double source representation found in Augusztinovicz (1999); Delillo et al. (2000); DeLillo et al. (2001; 2003); Raveendra et al. (1998); Schuhmacher et al. (2003); Tekatlian et al. (1996); Vlahopoulos & Raveerdra (1998); Williams et al. (2000); Zhang et al. (2001; 2000). Using the notation of the previous subsection the single source representation is defined as

$$p(\vec{x}) = (S_\Gamma \varphi)(\vec{x}), \quad \vec{x} \in \mathbb{R}^3, \tag{16}$$

where φ is the density function, the double source representation

$$p(\vec{x}) = (D_\Gamma \varphi)(\vec{x}), \quad \vec{x} \in \mathbb{R}^3 \backslash \Gamma. \tag{17}$$

and the combination of sources

$$p(\vec{x}) = \left(\left[S_\Gamma + i\eta D_\Gamma\right]\varphi\right)(\vec{x}), \quad \vec{x} \in \mathbb{R}^3 \backslash \Gamma, \tag{18}$$

where η is a constant.

2.3 Boundary maps

We define the well-known Dirichlet-to-Neumann γ_{dn} and Neumann-to-Dirichlet γ_{nd} boundary operators for $\vec{x} \in \Gamma$

$$p(\vec{x}) = (\gamma_{nd} v)^{\pm}(\vec{x}), \quad v(\vec{x}) = (\gamma_{dn} p)^{\pm}(\vec{x}). \tag{19}$$

For the direct formulation we found that

$$\begin{aligned}(\gamma_{nd}\phi)^{\pm}(\vec{x}) &:= i\rho\omega\left(\left[D_\Gamma^{\pm} \pm I\right]^{-1} S_\Gamma^{\pm}\phi\right)(\vec{x}),\\ (\gamma_{dn}\phi)^{\pm}(\vec{x}) &:= \frac{1}{i\rho\omega}\left(\left(S_\Gamma^{\pm}\right)^{-1}\left[D_\Gamma^{\pm} \pm I\right]\phi\right)(\vec{x}),\end{aligned} \tag{20}$$

where I is the identity operator. For the indirect formulation with single source formulation we have the relations

$$\begin{aligned}(\gamma_{nd}\phi)^{\pm}(\vec{x}) &:= i\rho\omega\left(S_\Gamma^{\pm}\left(K_\Gamma^{\pm}\right)^{-1}\phi\right)(\vec{x}),\\ (\gamma_{dn}\phi)^{\pm}(\vec{x}) &:= \frac{1}{i\rho\omega}\left(K_\Gamma^{\pm}\left(S_\Gamma^{\pm}\right)^{-1}\phi\right)(\vec{x}),\end{aligned} \tag{21}$$

for the indirect formulation with double source formulation

$$\begin{aligned}(\gamma_{nd}\phi)^{\pm}(\vec{x}) &:= i\rho\omega\left(D_\Gamma^{\pm}\left(H_\Gamma^{\pm}\right)^{-1}\phi\right)(\vec{x}),\\ (\gamma_{dn}\phi)^{\pm}(\vec{x}) &:= \frac{1}{i\rho\omega}\left(H_\Gamma^{\pm}\left(D_\Gamma^{\pm}\right)^{-1}\phi\right)(\vec{x}),\end{aligned} \tag{22}$$

and for the indirect formulation with a combination of sources

$$\begin{aligned}(\gamma_{nd}\phi)^{\pm}(\vec{x}) &:= i\rho\omega\left(\left(S_\Gamma^{\pm} + i\eta D_\Gamma^{\pm}\right)\left(K_\Gamma^{\pm} + i\eta H_\Gamma^{\pm}\right)^{-1}\phi\right)(\vec{x}),\\ (\gamma_{dn}\phi)^{\pm}(\vec{x}) &:= \frac{1}{i\rho\omega}\left(\left(K_\Gamma^{\pm} + i\eta H_\Gamma^{\pm}\right)\left(S_\Gamma^{\pm} + i\eta D_\Gamma^{\pm}\right)^{-1}\phi\right)(\vec{x}).\end{aligned} \tag{23}$$

2.4 Transfer functions

The traditional approach for solving NAH (both for direct and indirect formulations) is based on the operators

$$p(\vec{x}) = \left(\mathcal{G}_\Gamma^d p\right)(\vec{x}) := \int_\Gamma \Phi_d(\vec{x},\vec{y})p(\vec{y})dS(\vec{y}), \quad \vec{x} \in \mathbb{R}^3\backslash\Gamma, \tag{24}$$

$$p(\vec{x}) = \left(\mathcal{G}_\Gamma^n v\right)(\vec{x}) := \int_\Gamma \Phi_n(\vec{x},\vec{y})v(\vec{y})dS(\vec{y}), \quad \vec{x} \in \mathbb{R}^3\backslash\Gamma, \tag{25}$$

where the functions Φ_d and Φ_n are respectively the *Dirichlet Green's function* and *Neumann Green's function* for Equation 1. These integral equations are similar to the Rayleigh's integral formulas for planar NAH (see Williams (1999)). In general, the knowledge of an explicit

formula for these Green's functions will be helpful, but can only be found for some restricted geometries (see Ouellet et al. (1991); Pan & Bies (1990); Williams (1997)).
With the help of the Dirichlet-to-Neumann γ_{dn} and Neumann-to-Dirichlet γ_{nd} boundary operators we can find an explicit representation of the operators $\mathcal{G}^d_\Gamma, \mathcal{G}^n_\Gamma$ for the direct formulations

$$\begin{aligned} \left(\mathcal{G}^d_\Gamma \varphi\right)^{\pm}(\vec{x}) &:= \left(\pm \left[i\rho\omega S_\Gamma \left(\gamma^{\pm}_{nd}\right) - D_\Gamma\right] \varphi\right)(\vec{x}), \\ \left(\mathcal{G}^n_\Gamma \varphi\right)^{\pm}(\vec{x}) &:= \left(\pm \left[i\rho\omega S_\Gamma - D_\Gamma \left(\gamma^{\pm}_{dn}\right)\right] \varphi\right)(\vec{x}). \end{aligned} \tag{26}$$

The indirect formulations with single source will have the expressions

$$\begin{aligned} \left(\mathcal{G}^d_\Gamma \varphi\right)^{\pm}(\vec{x}) &:= \left(S_\Gamma \left(S^{\pm}_\Gamma\right)^{-1} \varphi\right)(\vec{x}), \\ \left(\mathcal{G}^n_\Gamma \varphi\right)^{\pm}(\vec{x}) &:= i\rho\omega \left(S_\Gamma \left(K^{\pm}_\Gamma\right)^{-1} \varphi\right)(\vec{x}). \end{aligned} \tag{27}$$

for the double source

$$\begin{aligned} \left(\mathcal{G}^d_\Gamma \varphi\right)^{\pm}(\vec{x}) &:= \left(D_\Gamma \left(D^{\pm}_\Gamma\right)^{-1} \varphi\right)(\vec{x}), \\ \left(\mathcal{G}^n_\Gamma \varphi\right)^{\pm}(\vec{x}) &:= i\rho\omega \left(D_\Gamma \left(H^{\pm}_\Gamma\right)^{-1} \varphi\right)(\vec{x}). \end{aligned} \tag{28}$$

and for a combination of sources

$$\begin{aligned} \left(\mathcal{G}^d_\Gamma \varphi\right)^{\pm}(\vec{x}) &:= \left(\left(S_\Gamma + i\eta D_\Gamma\right)\left(S^{\pm}_\Gamma + i\eta D^{\pm}_\Gamma\right)^{-1} \varphi\right)(\vec{x}), \\ \left(\mathcal{G}^n_\Gamma \varphi\right)^{\pm}(\vec{x}) &:= i\rho\omega \left(\left(S_\Gamma + i\eta D_\Gamma\right)\left(K^{\pm}_\Gamma + i\eta H^{\pm}_\Gamma\right)^{-1} \varphi\right)(\vec{x}). \end{aligned} \tag{29}$$

3. Numerical methods for integral equations

The numerical solution of an integral equation is based on its discretization, which is a reduction into a linear matrix system where numerical methods can be applied. Boundary Element Methods (BEM) have been successfully used in the area of acoustic radiation and scattering (see Bai (1992); Kang & Ih (2000a); Kim & Ih (1996); Langrenne & Garcia (1999); Seybert et al. (1985); Williams et al. (2000)) for three dimensional surfaces.

3.1 Boundary integral methods

The boundary elements used for approximating the surface integral are schematically shown in Fig. 2. N_v triangular or quadrilateral elements are used in this study for the construction of meshes. The global cartesian coordinates $\vec{y}$ on any point of an element $\triangle_j, j = 1, ..., N_v$ are assumed to be related to the nodes $\vec{y}_{(m,j)}, m = 1, .., n_q$ by

$$\vec{y}(\vec{\xi}) = \sum_{m=1}^{n_q} N_m(\vec{\xi})\vec{y}_{(m,j)} \tag{30}$$

in which $\vec{\xi} = (\xi_1, \xi_2)$ and shape functions $N_m(\vec{\xi})$ for the triangular elements are given in Table 1 and for quadrilateral elements are given in Table 2.
Equation 30 is an isoparametric transformation in which a surface element is mapped into a plane equilateral unit triangle or quadrilateral (as in the lower part of figure 2). Next, the

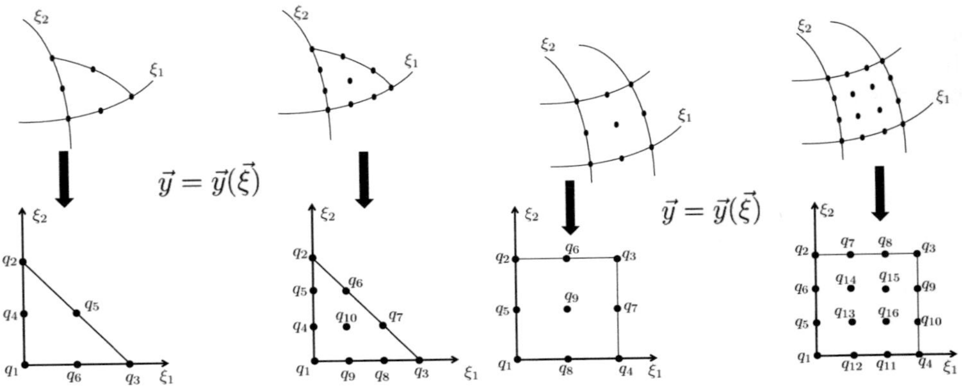

(a) triangular elements (b) quadrilateral elements

Fig. 2. Boundary elements used for iso-parametric transformation: linear, quadratic and cubic elements.

	Linear	Quadratic	Cubic
$N_1(\vec{\xi})$	u	$u(2u-1)$	$\frac{1}{2}u(3u-1)(3u-2)$
$N_2(\vec{\xi})$	ξ_2	$\xi_2(2\xi_2-1)$	$\frac{1}{2}\xi_2(3\xi_2-1)(3\xi_2-2)$
$N_3(\vec{\xi})$	ξ_1	$\xi_1(2\xi_1-1)$	$\frac{1}{2}\xi_1(3\xi_1-1)(3\xi_1-2)$
$N_4(\vec{\xi})$		$4\xi_2 u$	$\frac{9}{2}\xi_2(3u-1)u$
$N_5(\vec{\xi})$		$4\xi_1\xi_2$	$\frac{9}{2}\xi_2(3\xi_2-1)u$
$N_6(\vec{\xi})$		$4\xi_1 u$	$\frac{9}{2}\xi_1\xi_2(3\xi_2-1)$
$N_7(\vec{\xi})$			$\frac{9}{2}\xi_1(3\xi_1-1)\xi_2$
$N_8(\vec{\xi})$			$\frac{9}{2}\xi_1(3\xi_1-1)u$
$N_9(\vec{\xi})$			$\frac{9}{2}\xi_1(3u-1)u$
$N_{10}(\vec{\xi})$			$27\xi_1\xi_2 u$
	$u = 1-\xi_1-\xi_2$		
	$q_1=(0,0)$, $q_2=(0,1)$, $q_3=(1,0)$,	$q_4=(0,\frac{1}{2})$, $q_5=(\frac{1}{2},\frac{1}{2})$, $q_6=(\frac{1}{2},0)$	$q_4=(0,\frac{1}{3})$, $q_5=(0,\frac{2}{3})$, $q_6=(\frac{1}{3},\frac{2}{3})$, $q_7=(\frac{2}{3},\frac{1}{3})$, $q_8=(\frac{2}{3},0)$, $q_9=(\frac{1}{3},0)$, $q_{10}=(\frac{1}{3},\frac{1}{3})$,

Table 1. Shape functions for triangular elements.

boundary variable φ given in Equations 4-7 and Equations 9-12 will be represented on each element j according to

$$\varphi(\vec{\xi}) = \sum_{m=1}^{n_q} N_m(\vec{\xi})\varphi_{(m,j)} \tag{31}$$

where $\varphi_{(m,j)}$ are the values of φ at node m of the element j.

	Linear	Quadratic	Cubic
$N_1(\vec{\xi})$	$\frac{1}{4}(1-\xi_1)(1-\xi_2)$	$\frac{1}{4}\xi_1(\xi_1-1)\xi_2(\xi_2-1)$	$\frac{81}{256}(\xi_1^2-\frac{1}{9})(\xi_2^2-\frac{1}{9})(\xi_1-1)(\xi_2-1)$
$N_2(\vec{\xi})$	$\frac{1}{4}(1-\xi_1)(1+\xi_2)$	$\frac{1}{4}\xi_1(\xi_1-1)\xi_2(1+\xi_2)$	$\frac{81}{256}(\xi_1^2-\frac{1}{9})(\xi_2^2-\frac{1}{9})(1-\xi_1)(\xi_2+1)$
$N_3(\vec{\xi})$	$\frac{1}{4}(1+\xi_1)(1+\xi_2)$	$\frac{1}{4}\xi_1(1+\xi_1)\xi_2(1+\xi_2)$	$\frac{81}{256}(\xi_1^2-\frac{1}{9})(\xi_2^2-\frac{1}{9})(\xi_1+1)(\xi_2+1)$
$N_4(\vec{\xi})$	$\frac{1}{4}(1+\xi_1)(1-\xi_2)$	$\frac{1}{4}\xi_1(1+\xi_1)\xi_2(\xi_2-1)$	$\frac{81}{256}(\xi_1^2-\frac{1}{9})(\xi_2^2-\frac{1}{9})(\xi_1+1)(1-\xi_2)$
$N_5(\vec{\xi})$		$\frac{1}{2}\xi_1(\xi_1-1)(1-\xi_2^2)$	$\frac{243}{256}(\xi_1^2-\frac{1}{9})(\xi_2^2-1)(1-\xi_1)(\xi_2-\frac{1}{3})$
$N_6(\vec{\xi})$		$\frac{1}{2}(1-\xi_1^2)\xi_2(1+\xi_2)$	$\frac{243}{256}(\xi_1^2-\frac{1}{9})(\xi_2^2-1)(\xi_1-1)(\xi_2+\frac{1}{3})$
$N_7(\vec{\xi})$		$\frac{1}{2}\xi_1(1+\xi_1)(1-\xi_2^2)$	$\frac{243}{256}(\xi_1^2-1)(\xi_2^2-\frac{1}{9})(\xi_1-\frac{1}{3})(\xi_2+1)$
$N_8(\vec{\xi})$		$\frac{1}{2}(1-\xi_1^2)\xi_2(\xi_2-1)$	$\frac{243}{256}(1-\xi_1^2)(\xi_2^2-\frac{1}{9})(\xi_1+\frac{1}{3})(\xi_2+1)$
$N_9(\vec{\xi})$		$(1-\xi_1^2)(1-\xi_2^2)$	$\frac{243}{256}(\xi_1^2-\frac{1}{9})(1-\xi_2^2)(\xi_1+1)(\xi_2+\frac{1}{3})$
$N_{10}(\vec{\xi})$			$\frac{243}{256}(\xi_1^2-\frac{1}{9})(\xi_2^2-1)(\xi_1+1)(\xi_2-\frac{1}{3})$
$N_{11}(\vec{\xi})$			$\frac{243}{256}(\xi_1^2-1)(\xi_2^2-\frac{1}{9})(\xi_1+\frac{1}{3})(\xi_2-1)$
$N_{12}(\vec{\xi})$			$\frac{243}{256}(\xi_1^2-1)(\xi_2^2-\frac{1}{9})(\xi_1-\frac{1}{3})(1-\xi_2)$
$N_{13}(\vec{\xi})$			$\frac{729}{256}(\xi_1^2-1)(\xi_2^2-1)(\xi_1-\frac{1}{3})(\xi_2-\frac{1}{3})$
$N_{14}(\vec{\xi})$			$\frac{729}{256}(\xi_1^2-1)(\xi_2^2-1)(\frac{1}{3}-\xi_1)(\xi_2+\frac{1}{3})$
$N_{15}(\vec{\xi})$			$\frac{729}{256}(\xi_1^2-1)(\xi_2^2-1)(\xi_1+\frac{1}{3})(\xi_2+\frac{1}{3})$
$N_{16}(\vec{\xi})$			$\frac{729}{256}(\xi_1^2-1)(\xi_2^2-1)(\xi_1+\frac{1}{3})(\frac{1}{3}-\xi_2)$
	$q_1=(-1,-1)$, $q_2=(-1,1)$, $q_3=(1,1)$, $q_4=(1,-1)$	$q_5=(-1,0)$, $q_6=(0,1)$, $q_7=(1,0)$, $q_8=(0,-1)$, $q_9=(0,0)$	$q_5=(-1,-\frac{1}{3}, q_6=(-1,\frac{1}{3})$, $q_7=(-\frac{1}{3},1), q_8=(\frac{1}{3},1)$, $q_9=(1,\frac{1}{3}), q_{10}=(1,-\frac{1}{3})$, $q_{11}=(\frac{1}{3},-1), q_{12}=(-\frac{1}{3},-1)$, $q_{13}=(-\frac{1}{3},-\frac{1}{3}), q_{14}=(-\frac{1}{3},\frac{1}{3})$, $q_{15}=(\frac{1}{3},\frac{1}{3}), q_{16}=(\frac{1}{3},-\frac{1}{3})$

Table 2. Shape functions for quadrilateral elements.

For $\vec{x} \in \mathbb{R}^3 \backslash \Gamma$ the integrals in Equations 4-7 are approximated by

$$(S\varphi)(\vec{x}) \approx \sum_{j=1}^{N_v}\sum_{m=1}^{n_q} S^i_{(m,j)}\varphi_{(m,j)}, \quad (D\varphi)(\vec{x}) \approx \sum_{j=1}^{N_v}\sum_{m=1}^{n_q} D^i_{(m,j)}\varphi_{(m,j)}, \tag{32}$$

$$(K\varphi)(\vec{x}) \approx \sum_{j=1}^{N_v}\sum_{m=1}^{n_q} K^i_{(m,j)}\varphi_{(m,j)}, \quad (H\varphi)(\vec{x}) \approx \sum_{j=1}^{N_v}\sum_{m=1}^{n_q} H^i_{(m,j)}\varphi_{(m,j)},$$

where

$$S^i_{(m,j)} = \int_{\triangle} \Phi(\vec{x}_i,\vec{y}(\xi_1,\xi_2))N_m(\xi_1,\xi_2)J(\xi_1,\xi_2)d\xi_1 d\xi_2, \tag{33}$$

$$D^i_{(m,j)} = \int_{\triangle} \frac{\partial\Phi(\vec{x}_i,\vec{y}(\xi_1,\xi_2))}{\partial\vec{n}(\vec{y}(\xi_1,\xi_2))}N_m(\xi_1,\xi_2)J(\xi_1,\xi_2)d\xi_1 d\xi_2, \tag{34}$$

$$K^i_{(m,j)} = \int_{\triangle} \frac{\partial\Phi(\vec{x}_i,\vec{y}(\xi_1,\xi_2))}{\partial\vec{n}(\vec{x}_i)}N_m(\xi_1,\xi_2)J(\xi_1,\xi_2)d\xi_1 d\xi_2, \tag{35}$$

$$H^i_{(m,j)} = \int_{\triangle} \frac{\partial^2\Phi(\vec{x}_i,\vec{y}(\xi_1,\xi_2))}{\partial\vec{n}(\vec{y}(\xi_1,\xi_2))\partial\vec{n}(\vec{x}_i)}N_m(\xi_1,\xi_2)J(\xi_1,\xi_2)d\xi_1 d\xi_2, \tag{36}$$

in which $\triangle$ is the unit surface element, $J(\xi_1,\xi_2)$ is the Jacobian of the coordinate transformation. This Jacobian is known exactly for some surfaces, however for general surfaces it is more convenient to approximate it numerically using Equation 30

$$J(\xi_1,\xi_2) \approx |\partial_{\xi_1} N_m(\xi_1,\xi_2) \times \partial_{\xi_2} N_m(\xi_1,\xi_2)|$$

as suggested in Chien (1995). Integrals $S^i_{(m,j)}$, $D^i_{(m,j)}$, $K^i_{(m,j)}$, $H^i_{(m,j)}$ are approximated numerically using the adaptive integration scheme given in (Atkinson, 1997, Chapter 9) with the formula $(T_2 : 5.1)$ from (Strout, 1971, Chapter 8) for triangles and gaussian quadrature for quadrilaterals.
For $\vec{x} \in \Gamma$ the integrals in Equations 9-12 will be approximated by

$$(D\varphi)^{\pm}(\vec{x}) \approx \sum_{j=1}^{N_v} \sum_{m=1}^{q} D^i_{(m,j)} \varphi_{(m,j)} \pm \Omega(\vec{x})\varphi(\vec{x}), \quad (S\varphi)(\vec{x})^{\pm} \approx \sum_{j=1}^{N_v} \sum_{m=1}^{n_q} S^i_{(m,j)} \varphi_{(m,j)}, \tag{37}$$
$$(K\varphi)^{\pm}(\vec{x}) \approx \sum_{j=1}^{N_v} \sum_{m=1}^{q} K^i_{(m,j)} \varphi_{(m,j)} \mp \Omega(\vec{x})\varphi(\vec{x}), \quad (H\varphi)(\vec{x})^{\pm} \approx \sum_{j=1}^{N_v} \sum_{m=1}^{n_q} H^i_{(m,j)} \varphi_{(m,j)},$$

Integrals $S^i_{(m,j)}$, $D^i_{(m,j)}$, $K^i_{(m,j)}$ and $H^i_{(m,j)}$ will be approximated using the adaptive integration scheme used for the previous case when $\vec{x} \notin \triangle_j$. When $\vec{x} \in \triangle_j$ then $S^i_{(m,j)}$, $D^i_{(m,j)}$, and $K^i_{(m,j)}$ are approximated numerically using a Duffy transformation and Gaussian quadrature as suggested in Schwab & Wendland (1992).

3.2 Equivalent sources methods

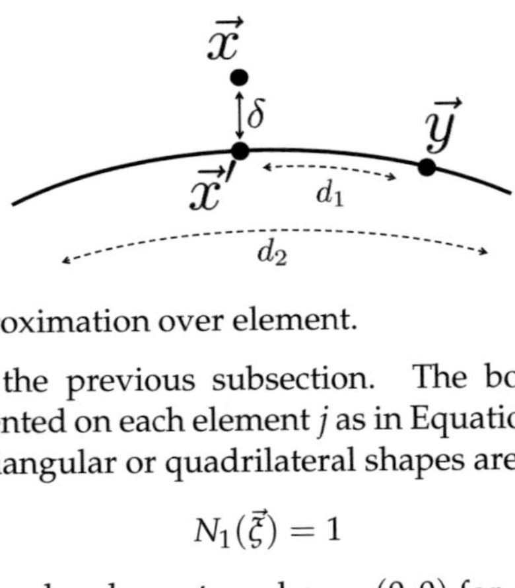

Fig. 3. Setup of integral approximation over element.

We utilize the notation of the previous subsection. The boundary variable φ given in Equations 4-7 will be represented on each element j as in Equation 31 using constant elements. The constant elements for triangular or quadrilateral shapes are

$$N_1(\vec{\xi}) = 1$$

and $q_1 = (1/3, 1/3)$ for triangular elements and $q_1 = (0,0)$ for quadrilateral elements.
For $\vec{x} \in \mathbb{R}^3 \backslash \Gamma$ the mean value theorem applied to Equation 32 gives

$$S^i_{(m,j)} = \Phi(\vec{x},\vec{\zeta})A_{\triangle}, \quad D^i_{(m,j)} = \frac{\partial\Phi(\vec{x},\vec{\zeta})}{\partial\vec{n}(\vec{\zeta})}A_{\triangle}, \tag{38}$$
$$K^i_{(m,j)} = \frac{\partial\Phi(\vec{x},\vec{\zeta})}{\partial\vec{n}(\vec{x})}A_{\triangle}, \quad D^i_{(m,j)} = \frac{\partial^2\Phi(\vec{x},\vec{\zeta})}{\partial\vec{n}(\vec{x})\partial\vec{n}(\vec{\zeta})}A_{\triangle},$$

for certain $\vec{\zeta} \in \triangle_j$ and

$$A_{\triangle_j} = \int_{\triangle} J(\xi_1,\xi_2)d\xi_1 d\xi_2,$$

The multi-dimensional Taylor series are used to expand around $\vec{y} \in \mathbb{R}^3$

$$\partial^{\alpha_1}_{\vec{n}(x)} \partial^{\alpha_2}_{\vec{n}(\vec{\zeta})} \Phi(\vec{x}, \vec{\zeta}) = \partial^{\alpha_1}_{\vec{n}(\vec{x})} \partial^{\alpha_2}_{\vec{n}(\vec{y})} \Phi(\vec{x}, \vec{y}) + D_1 + D_2 + \ldots \tag{39}$$

where $\alpha_1, \alpha_2 = 0, 1$ and

$$D_1 = \sum_{m=1}^{3} \partial_{y_m} \left(\partial^{\alpha_1}_{\vec{n}(x)} \partial^{\alpha_2}_{\vec{n}(\vec{y})} \Phi(\vec{x}, \vec{y}) \right) (\zeta_m - y_m),$$

$$D_2 = \sum_{m=1}^{3} \partial^2_{y_m} \left(\partial^{\alpha_1}_{\vec{n}(x)} \partial^{\alpha_2}_{\vec{n}(\vec{y})} \Phi(\vec{x}, \vec{y}) \right) \frac{(\zeta_m - y_m)^2}{2}$$

$$+ \sum_{m,n=1, m \neq n}^{3} \partial_{y_m} \partial_{y_n} \left(\partial^{\alpha_1}_{x_i} \partial^{\alpha_2}_{\vec{n}(\vec{y})} \Phi(\vec{x}, \vec{y}) \right) (\zeta_m - y_m)(\zeta_n - y_n).$$

From the derivatives of Φ we obtain the estimates

$$\left| \partial_{y_m} \partial^{\alpha_1}_{\vec{n}(\vec{x})} \partial^{\alpha_2}_{\vec{n}(\vec{y})} \Phi(\vec{x}, \vec{y}) \right| \leq \frac{A_{(1+\alpha_1+\alpha_2)}(k)}{4\pi |\vec{x} - \vec{y}|^{2+\alpha_1+\alpha_2}}, \quad \left| \partial_{y_m} \partial_{y_n} \partial^{\alpha_1}_{\vec{n}(\vec{x})} \partial^{\alpha_2}_{\vec{n}(\vec{y})} \Phi(\vec{x}, \vec{y}) \right| \leq \frac{A_{(2+\alpha_1+\alpha_2)}(k)}{4\pi |\vec{x} - \vec{y}|^{3+\alpha_1+\alpha_2}}$$

where

$$A_{(1)}(k) = C_1(k), \quad A_{(2)}(k) = C_1(k) + C_2(k), \quad A_{(3)}(k) = 3C_2(k) + C_3(k),$$
$$A_{(4)}(k) = 3C_2(k) + 4C_3(k) + C_4(k),$$
$$C_1(k) = |ik|\vec{x} - \vec{y}| - 1|,$$
$$C_2(k) = \left| k^2 |\vec{x} - \vec{y}|^2 + 3ik|\vec{x} - \vec{y}| - 3 \right|,$$
$$C_3(k) = \left| ik^3 |\vec{x} - \vec{y}|^3 - 6k^2 |\vec{x} - \vec{y}|^2 - 15ik|\vec{x} - \vec{y}| + 15 \right|,$$
$$C_4(k) = \left| k^4 |\vec{x} - \vec{y}|^4 + 10ik^3 |\vec{x} - \vec{y}|^3 - 45k^2 |\vec{x} - \vec{y}|^2 + 105ik|\vec{x} - \vec{y}| - 105 \right|,$$

We use the above estimates to obtain

$$|D_1| \leq \frac{3A_{(1+\alpha_1+\alpha_2)}(k)}{4\pi} \frac{|\vec{y} - \vec{\zeta}|}{|\vec{x} - \vec{y}|^{2+\alpha_1+\alpha_2}} = \frac{3A_{(1+\alpha_1+\alpha_2)}(k)}{4\pi} \frac{|\vec{y} - \vec{\zeta}|}{|\vec{x} - \vec{y}|^{1+\alpha_1+\alpha_2}} \left(\frac{|\vec{y} - \vec{\zeta}|}{|\vec{x} - \vec{y}|} \right),$$

$$|D_2| \leq \frac{9A_{(2+\alpha_1+\alpha_2)}(k)}{8\pi} \frac{|\vec{y} - \vec{\zeta}|^2}{|\vec{x} - \vec{y}|^{3+\alpha_1+\alpha_2}} = \frac{9A_{(2+\alpha_1+\alpha_2)}(k)}{8\pi} \frac{|\vec{y} - \vec{\zeta}|^2}{|\vec{x} - \vec{y}|^{1+\alpha_1+\alpha_2}} \left(\frac{|\vec{y} - \vec{\zeta}|}{|\vec{x} - \vec{y}|} \right)^2.$$

It is not hard to observe from Fig. 3 that $\delta \leq |\vec{x} - \vec{y}| \leq \delta + d_1$ and $|\vec{y} - \vec{\zeta}| < \sqrt{2} d_2$ where d_2 is the arc-length distance of the element $\triangle_j$ and d_1 the distance between $\vec{x}$ and $\vec{x}'$ which is the projection of $\vec{x}$ into the element $\triangle_j$. Under the assumption that $\delta > \sqrt{2} d_2$ then we can utilize the approximation

$$\partial^{\alpha_1}_{\vec{n}(x)} \partial^{\alpha_2}_{\vec{n}(\vec{\zeta})} \Phi(\vec{x}, \vec{\zeta}) = \partial^{\alpha_1}_{\vec{n}(\vec{x})} \partial^{\alpha_2}_{\vec{n}(\vec{y})} \Phi(\vec{x}, \vec{y}) + \frac{3A_{(1+\alpha_1+\alpha_2)}(k)}{4\pi} \frac{|\vec{y} - \vec{\zeta}|}{|\vec{x} - \vec{y}|^{2+\alpha_1+\alpha_2}} + O\left(\frac{|\vec{y} - \vec{\zeta}|}{|\vec{x} - \vec{y}|} \right)^2. \tag{40}$$

The error estimate in Equation 40 has been studied in Valdivia & Williams (2006) for the case $\alpha_1 = \alpha_2 = 0, \delta < d_2 < 2\delta$. A similar behavior of the error should be expected for other cases of α_1, α_2. Finally Equation 40 is used to justify the approximations of Equation 38 that results in the approximation

$$(S\varphi)(\vec{x}) \approx \sum_{j=1}^{N_v} \Phi(\vec{x}, \vec{y}_j) q_j, \tag{41}$$

$$(D\varphi)(\vec{x}) \approx \sum_{j=1}^{N_v} \frac{\partial \Phi(\vec{x}, \vec{y}_j)}{\partial \vec{n}(\vec{y}_j)} q_j, \tag{42}$$

$$(K\varphi)(\vec{x}) \approx \sum_{j=1}^{N_v} \frac{\partial \Phi(\vec{x}, \vec{y}_j)}{\partial \vec{n}(\vec{x})} q_j, \tag{43}$$

$$(H\varphi)(\vec{x}) \approx \sum_{j=1}^{N_v} \frac{\partial^2 \Phi(\vec{x}, \vec{y}_j)}{\partial \vec{n}(\vec{y}_j) \partial \vec{n}(\vec{x})} q_j, \tag{44}$$

where $q_j = \varphi_{(1,j)} S_{\triangle_j}$.

4. Near-field acoustic holography

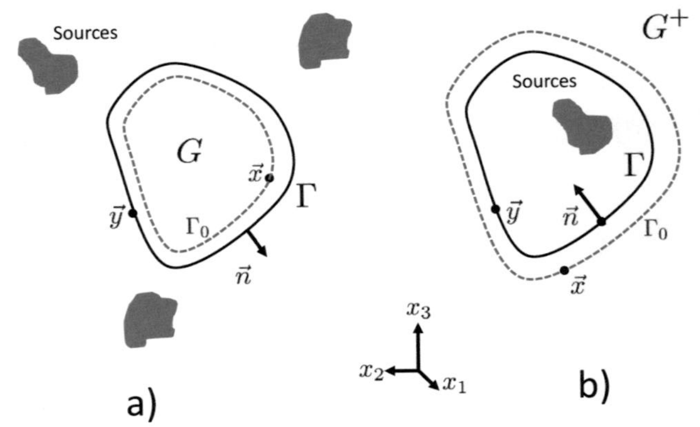

Fig. 4. Boundary setup for (a) interior NAH and (b) exterior NAH.

As shown in Fig. 4, for interior(exterior) NAH the acoustical sensors are placed on a surface Γ_0 inside (outside) the domain G. These are used to measure the pressure p and the fundamental problem is to recover the acoustic field (pressure, normal velocity and normal intensity) on Γ. The NAH problem for the *explicit approach* is reduced to the solution of the integral equations

$$\left(\mathcal{G}_\Gamma^d p\right)^{\pm}(\vec{x}) := p(\vec{x}), \quad \vec{x} \in \Gamma_0 \tag{45}$$

$$\left(\mathcal{G}_\Gamma^n v\right)^{\pm}(\vec{x}) := p(\vec{x}), \quad \vec{x} \in \Gamma_0. \tag{46}$$

The superscript sign "+" in Equations 45 and 46 is used for exterior NAH and "−" for interior NAH. This notation will be kept throughout the rest of this chapter.
In practice for the NAH technique we will take M pressure measurements on the surface Γ_0 and will want to recover N pressure or normal velocity points on the surface Γ. In general we will find that $M \geq N$, but there is no major theoretical contradiction if $M < N$. We will denote

as $\mathbf{p}$ the column vector with M pressure measurements on Γ_0. The N points of the recovered normal velocity and pressure on Γ will be given respectively by the column vectors $\mathbf{v}^s$, $\mathbf{p}^s$.

4.1 Boundary element methods

For each $\vec{x}_i \in \Gamma_0$ the sums in Equation 32 can be reduced for the N points on Γ to produce the $M \times M$ complex matrices $\left[\mathcal{S}_{(\Gamma_0,\Gamma)}\right]$, $\left[\mathcal{D}_{(\Gamma_0,\Gamma)}\right]$, $\left[\mathcal{K}_{(\Gamma_0,\Gamma)}\right]$ and $\left[\mathcal{H}_{(\Gamma_0,\Gamma)}\right]$ that correspond to discrete approximations of the operators in Equations 4-7. Similarly, for $\vec{x}_i \in \Gamma$ the sums in Equation 37 will be reduced to obtain the $N \times N$ complex matrices $\left[\mathcal{S}^{\pm}_{(\Gamma)}\right]$, $\left[\mathcal{D}^{\pm}_{(\Gamma)}\right]$, $\left[\mathcal{K}^{\pm}_{(\Gamma)}\right]$ and $\left[\mathcal{H}^{\pm}_{(\Gamma)}\right]$ that correspond to discrete approximations of the operators in Equations 9-12.

4.1.1 Explicit approach

The discretization of the Dirichlet-to-Neumann $\left[\mathbf{Y}^{\pm}_{(dn,\Gamma)}\right]$ and Neumann-to-Dirichlet $\left[\mathbf{Y}^{\pm}_{(nd,\Gamma)}\right]$ boundary operators gives the matrix equation

$$\mathbf{v}^s = \left[\mathbf{Y}^{\pm}_{(dn,\Gamma)}\right]\mathbf{p}^s, \quad \mathbf{p}^s = \left[\mathbf{Y}^{\pm}_{(nd,\Gamma)}\right]\mathbf{v}^s \tag{47}$$

For the direct formulation we found that

$$\begin{aligned}\left[\mathbf{Y}^{\pm}_{(nd,\Gamma)}\right] &:= i\rho\omega\left(\left[\mathcal{D}^{\pm}_{(\Gamma)}\right] \pm [\mathbf{I}]\right)^{-1}\left[\mathcal{S}^{\pm}_{(\Gamma)}\right],\\ \left[\mathbf{Y}^{\pm}_{(dn,\Gamma)}\right] &:= \frac{1}{i\rho\omega}\left[\mathcal{S}^{\pm}_{(\Gamma)}\right]^{-1}\left(\left[\mathcal{D}^{\pm}_{(\Gamma)}\right] \pm [\mathbf{I}]\right),\end{aligned} \tag{48}$$

where $[\mathbf{I}]$ is the $N \times N$ identity matrix. For the indirect formulation with single source formulation we have the relations

$$\begin{aligned}\left[\mathbf{Y}^{\pm}_{(nd,\Gamma)}\right] &:= i\rho\omega\left[\mathcal{S}^{\pm}_{(\Gamma)}\right]\left[\mathcal{K}^{\pm}_{(\Gamma)}\right]^{-1},\\ \left[\mathbf{Y}^{\pm}_{(dn,\Gamma)}\right] &:= \frac{1}{i\rho\omega}\left[\mathcal{K}^{\pm}_{(\Gamma)}\right]\left[\mathcal{S}^{\pm}_{(\Gamma)}\right]^{-1},\end{aligned} \tag{49}$$

for the indirect formulation with double source formulation

$$\begin{aligned}\left[\mathbf{Y}^{\pm}_{(nd,\Gamma)}\right] &:= i\rho\omega\left[\mathcal{D}^{\pm}_{(\Gamma)}\right]\left[\mathcal{H}^{\pm}_{(\Gamma)}\right]^{-1},\\ \left[\mathbf{Y}^{\pm}_{(dn,\Gamma)}\right] &:= \frac{1}{i\rho\omega}\left[\mathcal{H}^{\pm}_{(\Gamma)}\right]\left[\mathcal{D}^{\pm}_{(\Gamma)}\right]^{-1},\end{aligned} \tag{50}$$

and for the indirect formulation with a combination of sources

$$\begin{aligned}\left[\mathbf{Y}^{\pm}_{(nd,\Gamma)}\right] &:= i\rho\omega\left(\left[\mathcal{S}^{\pm}_{(\Gamma)}\right] + i\eta\left[\mathcal{D}^{\pm}_{(\Gamma)}\right]\right)\left(\left[\mathcal{K}^{\pm}_{(\Gamma)}\right] + i\eta\left[\mathcal{H}^{\pm}_{(\Gamma)}\right]\right)^{-1},\\ \left[\mathbf{Y}^{\pm}_{(dn,\Gamma)}\right] &:= \frac{1}{i\rho\omega}\left(\left[\mathcal{K}^{\pm}_{(\Gamma)}\right] + i\eta\left[\mathcal{H}^{\pm}_{(\Gamma)}\right]\right)\left(\left[\mathcal{S}^{\pm}_{(\Gamma)}\right] + i\eta\left[\mathcal{D}^{\pm}_{(\Gamma)}\right]\right)^{-1}.\end{aligned} \tag{51}$$

The boundary element methods give the discretization of Equations 45 and 46

$$\left[\mathcal{G}^{(d,\pm)}_{(\Gamma_0,\Gamma)}\right]\mathbf{p}^s = \mathbf{p}, \quad \left[\mathcal{G}^{(n,\pm)}_{(\Gamma_0,\Gamma)}\right]\mathbf{v}^s = \mathbf{p}. \tag{52}$$

With the help of the boundary operators $\left[\mathbf{Y}^{\pm}_{(dn,\Gamma)}\right]$, $\left[\mathbf{Y}^{\pm}_{(nd,\Gamma)}\right]$ we can find an explicit representation of the operators $\left[\mathcal{G}^{(d,\pm)}_{(\Gamma_0,\Gamma)}\right]$, $\left[\mathcal{G}^{(n,\pm)}_{(\Gamma_0,\Gamma)}\right]$ for the direct formulations

$$\begin{aligned}\left[\mathcal{G}^{(d,\pm)}_{(\Gamma_0,\Gamma)}\right] &:= \pm\left(i\rho\omega\left[\mathcal{S}_{(\Gamma_0,\Gamma)}\right]\left[\mathbf{Y}^{\pm}_{(nd,\Gamma)}\right] - \left[\mathcal{D}_{(\Gamma_0,\Gamma)}\right]\right),\\ \left[\mathcal{G}^{(n,\pm)}_{(\Gamma_0,\Gamma)}\right] &:= \pm\left(i\rho\omega\left[\mathcal{S}_{(\Gamma_0,\Gamma)}\right] - \left[\mathcal{D}_{(\Gamma_0,\Gamma)}\right]\left[\mathbf{Y}^{\pm}_{(dn,\Gamma)}\right]\right).\end{aligned} \tag{53}$$

The indirect formulations with single source will have the expressions

$$\begin{aligned}\left[\mathcal{G}^{(d,\pm)}_{(\Gamma_0,\Gamma)}\right] &:= \left[\mathcal{S}_{(\Gamma_0,\Gamma)}\right]\left[\mathcal{S}^{\pm}_{(\Gamma)}\right]^{-1},\\ \left[\mathcal{G}^{(n,\pm)}_{(\Gamma_0,\Gamma)}\right] &:= i\rho\omega\left[\mathcal{S}_{(\Gamma_0,\Gamma)}\right]\left[\mathcal{K}^{\pm}_{(\Gamma)}\right]^{-1},\end{aligned} \tag{54}$$

for the double source

$$\begin{aligned}\left[\mathcal{G}^{(d,\pm)}_{(\Gamma_0,\Gamma)}\right] &:= \left[\mathcal{D}_{(\Gamma_0,\Gamma)}\right]\left[\mathcal{D}^{\pm}_{(\Gamma)}\right]^{-1},\\ \left[\mathcal{G}^{(n,\pm)}_{(\Gamma_0,\Gamma)}\right] &:= i\rho\omega\left[\mathcal{D}_{(\Gamma_0,\Gamma)}\right]\left[\mathcal{H}^{\pm}_{(\Gamma)}\right]^{-1},\end{aligned} \tag{55}$$

and for a combination of sources

$$\begin{aligned}\left[\mathcal{G}^{(d,\pm)}_{(\Gamma_0,\Gamma)}\right] &:= \left(\left[\mathcal{S}_{(\Gamma_0,\Gamma)}\right] + i\eta\left[\mathcal{D}_{(\Gamma_0,\Gamma)}\right]\right)\left(\left[\mathcal{S}^{\pm}_{(\Gamma)}\right] + i\eta\left[\mathcal{H}^{\pm}_{(\Gamma)}\right]\right)^{-1},\\ \left[\mathcal{G}^{(n,\pm)}_{(\Gamma_0,\Gamma)}\right] &:= i\rho\omega\left(\left[\mathcal{S}_{(\Gamma_0,\Gamma)}\right] + i\eta\left[\mathcal{D}_{(\Gamma_0,\Gamma)}\right]\right)\left(\left[\mathcal{K}^{\pm}_{(\Gamma)}\right] + i\eta\left[\mathcal{H}^{\pm}_{(\Gamma)}\right]\right)^{-1}.\end{aligned} \tag{56}$$

4.1.2 Implicit approach

Using the boundary element method for the indirect formulation we obtain the equations for the single source

$$\left[\mathcal{S}_{(\Gamma_0,\Gamma)}\right]\varphi = \mathbf{p}, \tag{57}$$

$$\left[\mathcal{S}^{\pm}_{(\Gamma)}\right]\varphi = \mathbf{p}^s, \quad \frac{1}{i\rho\omega}\left[\mathcal{K}^{\pm}_{(\Gamma)}\right]\varphi = \mathbf{v}^s, \tag{58}$$

for the double source

$$\left[\mathcal{D}_{(\Gamma_0,\Gamma)}\right]\varphi = \mathbf{p}, \tag{59}$$

$$\left[\mathcal{D}^{\pm}_{(\Gamma)}\right]\varphi = \mathbf{p}^s, \quad \frac{1}{i\rho\omega}\left[\mathcal{H}^{\pm}_{(\Gamma)}\right]\varphi = \mathbf{v}^s, \tag{60}$$

for a combination of sources

$$\left(\left[\mathcal{S}_{(\Gamma_0,\Gamma)}\right] + i\eta\left[\mathcal{D}_{(\Gamma_0,\Gamma)}\right]\right)\varphi = \mathbf{p}, \tag{61}$$

$$\left(\left[\mathcal{S}^{\pm}_{(\Gamma)}\right] + i\eta\left[\mathcal{D}^{\pm}_{(\Gamma)}\right]\right)\varphi = \mathbf{p}^s, \quad \frac{1}{i\rho\omega}\left(\left[\mathcal{K}^{\pm}_{(\Gamma)}\right] + i\eta\left[\mathcal{H}^{\pm}_{(\Gamma)}\right]\right)\varphi = \mathbf{v}^s, \tag{62}$$

and for the direct approach the system

$$\begin{bmatrix}\mathbf{p}\\ 0\end{bmatrix} = \begin{bmatrix}\pm i\rho\omega\left[\mathcal{S}_{(\Gamma_0,\Gamma)}\right] & \mp\left[\mathcal{D}_{(\Gamma_0,\Gamma)}\right]\\ \pm i\rho\omega\left[\mathcal{S}^{\pm}_{(\Gamma)}\right] & \mp\left(\left[\mathcal{D}^{\pm}_{(\Gamma)}\right] \pm [\mathbf{I}]\right)\end{bmatrix}\begin{bmatrix}\mathbf{v}^s\\ \mathbf{p}^s\end{bmatrix} \tag{63}$$

4.1.3 Modified approach

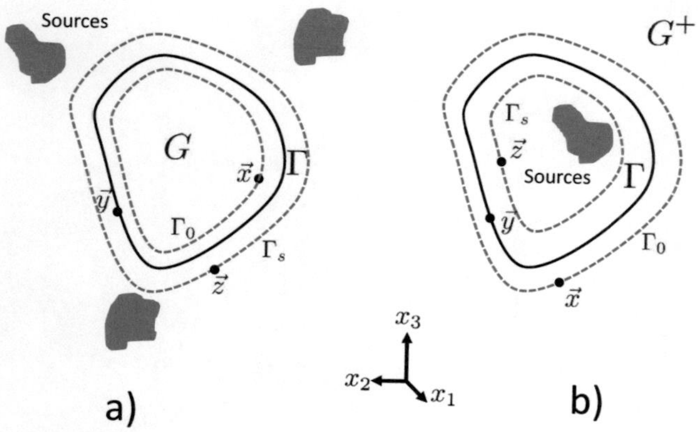

Fig. 5. Boundary setup with source surface for (a) interior NAH and (b) exterior NAH.

One of the major drawbacks of the previous methods is the calculation of the singularity for the matrices $\left[\mathcal{S}^{\pm}_{(\Gamma)}\right]$,$\left[\left[\mathcal{D}^{\pm}_{(\Gamma)}\right]\right]$, $\left[\mathcal{K}^{\pm}_{(\Gamma)}\right]$ and $\left[\mathcal{H}^{\pm}_{(\Gamma)}\right]$. A modified approach requires the definition of a conformal surface Γ_s as in Fig. 4. We construct the $M \times N_s$ complex matrices $\left[\mathcal{S}_{(\Gamma_0,\Gamma_s)}\right]$,$\left[\mathcal{D}_{(\Gamma_0,\Gamma_s)}\right]$, $\left[\mathcal{K}_{(\Gamma_0,\Gamma_s)}\right]$ and $\left[\mathcal{H}_{(\Gamma_0,\Gamma_s)}\right]$ in the same way as in previous sections, but the integration is over N_s points in the surface Γ_s. Similarly we obtain the $N \times N_s$ complex matrices $\left[\mathcal{S}_{(\Gamma,\Gamma_s)}\right]$,$\left[\mathcal{D}_{(\Gamma,\Gamma_s)}\right]$, $\left[\mathcal{K}_{(\Gamma,\Gamma_s)}\right]$ and $\left[\mathcal{H}_{(\Gamma,\Gamma_s)}\right]$ that correspond to discrete approximations of the operators in Equations 9-12.

For the indirect formulation we obtain the equations for the single source

$$\left[\mathcal{S}_{(\Gamma_0,\Gamma_s)}\right]\boldsymbol{\varphi} = \mathbf{p}, \tag{64}$$

$$\left[\mathcal{S}_{(\Gamma,\Gamma_s)}\right]\boldsymbol{\varphi} = \mathbf{p}^s, \quad \frac{1}{i\rho\omega}\left[\mathcal{K}_{(\Gamma,\Gamma_s)}\right]\boldsymbol{\varphi} = \mathbf{v}^s, \tag{65}$$

for the double source

$$\left[\mathcal{D}_{(\Gamma_0,\Gamma_s)}\right]\boldsymbol{\varphi} = \mathbf{p}, \tag{66}$$

$$\left[\mathcal{D}_{(\Gamma,\Gamma_s)}\right]\boldsymbol{\varphi} = \mathbf{p}^s, \quad \frac{1}{i\rho\omega}\left[\mathcal{H}_{(\Gamma,\Gamma_s)}\right]\boldsymbol{\varphi} = \mathbf{v}^s, \tag{67}$$

and for a combination of sources

$$\left(\left[\mathcal{S}_{(\Gamma_0,\Gamma_s)}\right] + i\eta\left[\mathcal{D}_{(\Gamma_0,\Gamma_s)}\right]\right)\boldsymbol{\varphi} = \mathbf{p}, \tag{68}$$

$$\left(\left[\mathcal{S}_{(\Gamma,\Gamma_s)}\right] + i\eta\left[\mathcal{D}_{(\Gamma,\Gamma_s)}\right]\right)\boldsymbol{\varphi} = \mathbf{p}^s, \quad \frac{1}{i\rho\omega}\left(\left[\mathcal{D}_{(\Gamma,\Gamma_s)}\right] + i\eta\left[\mathcal{H}_{(\Gamma,\Gamma_s)}\right]\right)\boldsymbol{\varphi} = \mathbf{v}^s. \tag{69}$$

4.2 Equivalent sources method

The equivalent sources method can be understood as an approximation of the boundary element method modified approach. For $\vec{x}_i, i = 1, ..., M$, in Γ_0 and $\vec{z}_i, i = 1, ..., N_s$ in Γ_s we obtain the matrix approximations of $\left[\boldsymbol{S}_{(\Gamma,\Gamma_s)}\right]$, $\left[\boldsymbol{D}_{(\Gamma,\Gamma_s)}\right]$, $\left[\boldsymbol{K}_{(\Gamma,\Gamma_s)}\right]$ and $\left[\boldsymbol{H}_{(\Gamma,\Gamma_s)}\right]$ given by Equation 41-44. There

$$\left[\boldsymbol{S}_{(\Gamma_0,\Gamma_s)}\right]\mathbf{q} = \mathbf{p}, \tag{70}$$

$$\left[\boldsymbol{S}_{(\Gamma,\Gamma_s)}\right]\mathbf{q} = \mathbf{p}^s, \quad \frac{1}{i\rho\omega}\left[\boldsymbol{K}_{(\Gamma,\Gamma_s)}\right]\mathbf{q} = \mathbf{v}^s, \tag{71}$$

for the double source

$$\left[\boldsymbol{D}_{(\Gamma_0,\Gamma_s)}\right]\mathbf{q} = \mathbf{p}, \tag{72}$$

$$\left[\boldsymbol{D}_{(\Gamma,\Gamma_s)}\right]\mathbf{q} = \mathbf{p}^s, \quad \frac{1}{i\rho\omega}\left[\boldsymbol{H}_{(\Gamma,\Gamma_s)}\right]\mathbf{q} = \mathbf{v}^s, \tag{73}$$

and for a combination of sources

$$\left(\left[\boldsymbol{S}_{(\Gamma_0,\Gamma_s)}\right] + i\eta\left[\boldsymbol{D}_{(\Gamma_0,\Gamma_s)}\right]\right)\mathbf{q} = \mathbf{p}, \tag{74}$$

$$\left(\left[\boldsymbol{S}_{(\Gamma,\Gamma_s)}\right] + i\eta\left[\boldsymbol{D}_{(\Gamma,\Gamma_s)}\right]\right)\mathbf{q} = \mathbf{p}^s, \quad \frac{1}{i\rho\omega}\left(\left[\boldsymbol{K}_{(\Gamma,\Gamma_s)}\right] + i\eta\left[\boldsymbol{H}_{(\Gamma,\Gamma_s)}\right]\right)\mathbf{q} = \mathbf{v}^s. \tag{75}$$

5. Numerical regularization

For the experimental problem, the exact pressure $\mathbf{p}$ is perturbed by measurement errors. We denote the measured pressure as $\widetilde{\mathbf{p}}$. If the elements of the perturbation $\mathbf{e} = \widetilde{\mathbf{p}} - \mathbf{p}$ are Gaussian (unbiased and uncorrelated) with covariance matrix $\sigma_0^2[\mathbf{I}]$, then $E\left(\|\mathbf{e}\|_2^2\right) = M\sigma_0^2$, where $\|.\|_2$ is the 2-norm. It is well known that the linear systems in Equation 52, Equations 57,59,61,63, Equations 64, 66,68, and Equations 70, 72,74, are ill-posed, i.e., the errors in $\widetilde{\mathbf{p}}$ will be amplified on the solutions $\mathbf{p}^s$, $\mathbf{v}^s$, $\boldsymbol{\varphi}$ or $\mathbf{q}$ and in most of the cases the recovery will be useless.
Consider the solution of the generic ill-posed linear matrix system

$$[\mathbf{A}]\,\mathbf{z} = \widetilde{\mathbf{p}}. \tag{76}$$

Here $[\mathbf{A}]$, $\mathbf{z}$, $\widetilde{\mathbf{p}}$ represent the ill-posed matrix, the solution of the linear system and the measurement vector of Equations 52,57,59, 61,63, 64, 66, 68, 70, 72, and 74. Let $\overline{M} \times \overline{N}$ be the dimension of the matrix $[\mathbf{A}]$. Then $\mathbf{z}$, $\widetilde{\mathbf{P}}$ are column vectors of $\overline{N}$, $\overline{M}$ entries respectively. For Equations 52,57,59, and 61, $\overline{M} = M$ and $\overline{N} = N$. For Equation 63, $\overline{M} = 2M$ and $\overline{N} = 2N$. For Equations 64,66,68, 70, 72, and 74, $\overline{M} = M$ and $\overline{N} = N_s$. The ill-posedness of the matrix system Equation 76 can be explained using the singular value decomposition (SVD)

$$[\mathbf{A}] = [\mathbf{U}][\boldsymbol{\Sigma}][\mathbf{V}]^H$$

where $[\mathbf{U}]$ and $[\mathbf{V}]$ are unitary matrices of dimension $\overline{M} \times \overline{M}$ and $\overline{N} \times \overline{N}$, respectively, and $[\mathbf{V}]^H$ is the conjugate transpose of $[\mathbf{V}]$. $[\boldsymbol{\Sigma}]$ = is a diagonal matrix with values $\sigma_1 \geq ... \geq \sigma_{N^*} \geq 0$ where $N^* = \text{rank}\,([\mathbf{A}])$. The values σ_i are called the singular values.
Denote $\mathbf{u}_i$, $i = 1, ..., M$ and $\mathbf{v}_i$, $i = 1, ..., N$, the columns of $[\mathbf{U}]$ and $[\mathbf{V}]$ respectively. For NAH, $\mathbf{u}_i$, $\mathbf{v}_i$ will be an approximation to the acoustic field by basic acoustic waves or *mode shapes* (see Williams et al. (2000)). In particular when $[\mathbf{A}] = \left[\mathcal{G}^{(n,\pm)}_{(\Gamma_0,\Gamma)}\right]$ (as in Equation 52), $\mathbf{u}_i$ will be the modes of the measured pressure on Γ_0 and $\mathbf{v}_i$ will be the modes of the normal velocity

on Γ. These modes are organized in such a way that the first modes are related with the non-evanescent waves and the last modes will be related with the evanescent waves.
The solution to Equation 76 by conventional methods can be explicitly written as

$$\mathbf{z}_{LS} = \sum_{i=1}^{N^*} \frac{\left(\mathbf{u}_i^H \widetilde{\mathbf{p}}\right)}{\sigma_i} \mathbf{v}_i. \tag{77}$$

Notice that this solution will include modes $\mathbf{u}_i$, $\mathbf{v}_i$ that are related to both non-evanescent and evanescent waves. It is well known that the modes related to the evanescent waves will produce the fine details of the reconstruction $\mathbf{z}_{LS}$, but at the same time will amplify the noise in $\widetilde{\mathbf{p}}$. Regularization methods for the solution to this inverse acoustic problem will need to include enough of these modes in order to obtain the desired resolution, and at the same time exclude some of these modes that are totally corrupted with noise.
As explained in Hansen (1998), regularization methods for the solution $\mathbf{z}$ of Equation 76 can be distinguished as *direct* and *iterative*.

5.1 Direct regularization

For the direct regularization the solution of Equation 76 is written explicitly with the help of the SVD as

$$\mathbf{z}_\alpha = \sum_{i=1}^{N^*} f_i^\alpha \left(\frac{\mathbf{u}_i^H \widetilde{\mathbf{p}}}{\sigma_i}\right) \mathbf{v}_i, \tag{78}$$

where α is the regularization parameter and f_i^α, $i = 1, ..., N^*$ are the *filter factors* (see Engl et al. (1996); Hansen (1998)). The parameter α will control the inclusion of modes related to evanescent waves into the solution $\mathbf{z}_\alpha$. The topic of direct regularization methods for linear matrix systems has been extensively studied in the last few decades. The best known regularization approach is the classical Tikhonov regularization, but there are many other approaches depending on the particular problem like damped SVD, truncated SVD or Tikhonov with High-pass filter. These approaches are defined by their respective filters

$$f^{(1,\alpha_i)} = \frac{\sigma_i^2}{\sigma_i^2 + \alpha^2}, \quad f^{(2,\alpha_i)} = \frac{\sigma_i}{\sigma_i + \alpha}, \quad f^{(3,\alpha_i)} = \begin{cases} 1, i \leq \alpha \\ 0, i > \alpha \end{cases}, \quad f^{(4,\alpha_i)} = \frac{\sigma_i^2}{\sigma_i^2 + \alpha^2 \left(\frac{\alpha^2}{\alpha^2 + \sigma_i^2}\right)}.$$

Tikhonov regularization with a high-pass filter (see Williams (2001)) uses the physical behavior of the evanescent waves to produce an optimal filter.

5.2 Iterative methods

Iterative regularization methods for the linear system Equation 76 are based on iteration schemes that access the coefficient matrix $[\mathbf{A}]$ only via matrix-vector multiplications with $[\mathbf{A}]$ and $[\mathbf{A}]^H$. They produce a sequence of iteration vectors $\mathbf{z}_{(l)}$, $l = 1, 2, 3, ...$, that converge to $\vec{z}_{LS}$ in Equation 78. For ill-posed linear systems these methods produce the phenomena of "semi-convergence", i.e., the vector $\mathbf{z}_{(l)}$ approaches the optimal regularization solution after a few iterations l. If the iteration is not stopped, the method converges to the least squares solution $\mathbf{z}_{LS}$ in Equation 78 which is generally totally corrupted by the noisy data $\widetilde{\mathbf{p}}$. In this case each iteration vector $\mathbf{z}_{(l)}$ can be considered as a regularized solution, with the iteration number l playing the role of the regularization parameter. These iterative methods are preferable to direct methods when the matrix $[\mathbf{A}]$ is so large that it is too time consuming or too memory-demanding to work with the SVD of $[\mathbf{A}]$. In the next section we will discuss a special type of iterative methods named Krylov subspace methods.

$$
\begin{array}{l}
\text{starting vector } \mathbf{z}_{(0)} = 0 \\
\mathbf{r}_{(0)} = \widetilde{\mathbf{p}} \\
\mathbf{d}_{(0)} = [\mathbf{A}]^H \mathbf{r}_{(0)} \\
\text{for } l = 1, 2, \dots (\leq \min(M, N)) \\
\quad \alpha_{(l)} = \left\| [\mathbf{A}]^H \mathbf{r}_{(l-1)} \right\|_2^2 / \left\| [\mathbf{A}] \, \mathbf{d}_{(l-1)} \right\|_2^2, \\
\quad \mathbf{z}_{(l)} = \mathbf{z}_{(l-1)} + \alpha_{(l)} \mathbf{d}_{(l-1)}, \\
\quad \mathbf{r}_{(l)} = \mathbf{r}_{(l-1)} - \alpha_{(l)} [\mathbf{A}] \, \mathbf{d}_{(l-1)} \\
\quad \beta_{(l)} = \left\| [\mathbf{A}]^H \mathbf{r}_{(l)} \right\|_2^2 / \left\| [\mathbf{A}]^H \mathbf{r}_{(l-1)} \right\|_2^2, \\
\quad \mathbf{d}_{(l)} = [\mathbf{A}]^H \mathbf{r}_{(l)} + \beta_{(l)} \mathbf{d}_{(l-1)},
\end{array}
$$

Table 3. Algorithm for Conjugate Gradients for the Normal Equations.

$$
\begin{array}{l}
\text{initial step} \\
\mathbf{z}_{(0)} = 0 \\
\beta_{(1)} \mathbf{u}_{(1)} = \widetilde{\mathbf{p}}, \quad \alpha_{(1)} \mathbf{v}_{(1)} = [\mathbf{A}]^H \mathbf{u}_{(1)} \\
\mathbf{w}_{(1)} = \mathbf{v}_{(1)}, \quad \overline{\phi}_{(1)} = \beta_{(1)}, \overline{\rho}_{(1)} = \alpha_{(1)} \\
\text{for } l = 1, 2, \dots (\leq N^*) \\
\quad \mathbf{p} = [\mathbf{A}] \mathbf{v}_{(l)} - \alpha_{(l)} \mathbf{u}_{(l)} \\
\quad \left.\begin{array}{l} \text{for } j = 1, \dots, (l-1) \\ \quad \mathbf{p} = \mathbf{p} - \left(\mathbf{u}_{(j)}^H \mathbf{p} \right) \mathbf{u}_{(j)} \end{array}\right\} \text{(Reorthogonalization)} \\
\quad \beta_{(l+1)} \mathbf{u}_{(l+1)} = \mathbf{p} \\
\quad \mathbf{r} = [\mathbf{A}]^H \mathbf{u}_{(l)} - \beta_{(l)} \mathbf{v}_{(l-1)} \\
\quad \left.\begin{array}{l} \text{for } j = 1, \dots, (l-1) \\ \quad \mathbf{r} = \mathbf{r} - \left(\mathbf{v}_{(j)}^H \mathbf{r} \right) \mathbf{v}_{(j)} \end{array}\right\} \text{(Reorthogonalization)} \\
\quad \alpha_{(l+1)} \mathbf{v}_{(l+1)} = \mathbf{r}_{(l)} \\
\quad \left.\begin{array}{l} \rho_{(l)} = \sqrt{(\overline{\rho}_{(l)})^2 + \beta_{(l+1)}^2} \\ c_{(l)} = \overline{\rho}_{(l)} / \rho_{(l)} \quad s_{(l)} = \beta_{(l+1)} / \rho_{(l)} \end{array}\right\} \text{(Givens Rotations)} \\
\quad \theta_{(l+1)} = s_{(l)} \alpha_{(l+1)}, \quad \overline{\rho}_{(l+1)} = c_{(l)} \alpha_{(l+1)} \\
\quad \phi_{(l+1)} = c_{(l)} \overline{\phi}_{(l)}, \quad \overline{\phi}_{(l+1)} = -s_{(l)} \overline{\phi}_{(l)} \\
\quad \mathbf{z}_{(l)} = \mathbf{z}_{(l-1)} + \left(\frac{\phi_{(l)}}{\rho_{(l)}} \right) \mathbf{w}_{(l)} \\
\quad \mathbf{w}_{(l+1)} = \mathbf{v}_{(l+1)} - \left(\frac{\theta_{(l+1)}}{\rho_{(l)}} \right) \mathbf{w}_{(l)}
\end{array}
$$

Table 4. Algorithm for LSQR.

5.2.1 Krylov subpace iterative methods

This section presents some general features of the Krylov subspace methods, and focuses on two of these methods: conjugate gradients for the normal equations (CGNE) (see Table 3) and least squares QR (LSQR) (see Table 4). Consider the normal equations of Equation 76,

$$[\mathbf{A}]^H [\mathbf{A}] \mathbf{z} = [\mathbf{A}]^H \widetilde{\mathbf{p}}. \tag{79}$$

The solution to Equation 79 is equal to the least squares solution $\mathbf{z}_{LS}$ given in Equation 78. The CGNE iterative algorithm (given in Table 3) produces, for each iteration l, the vector $\mathbf{z}_{(l)}$ that approximates the least squares solution $\mathbf{z}_{LS}$ to Equation 79. Notice from Table 3, line 6, that on iteration $l = 1$ of CGNE, $\mathbf{z}_{(1)}$ is formed by the vector $[\mathbf{A}]^H \widetilde{\mathbf{p}}$ multiplied by a constant. On iteration $l = 2$, $\mathbf{z}_{(2)}$ is formed by $c_1 [\mathbf{A}]^H \widetilde{\mathbf{p}} + c_2 \left([\mathbf{A}]^H [\mathbf{A}]\right) [\mathbf{A}]^H \widetilde{\mathbf{p}}$, where c_1, c_2 are constants. Similarly, for each l, a vector $\mathbf{z}_{(l)}$ can be expressed as

$$\mathbf{z}_{(l)} = \sum_{i=0}^{l-1} c_i \left([\mathbf{A}]^H [\mathbf{A}]\right)^i [\mathbf{A}]^H \widetilde{\mathbf{p}}, \tag{80}$$

where c_i, $i = 0, ..., l-1$ are constants. In mathematical notation, this combination of vectors is called the *Krylov subspace* l, and is denoted as $\mathcal{K}^{(l)}\left([\mathbf{A}]^H \widetilde{\mathbf{p}}, [\mathbf{A}]^H [\mathbf{A}]\right)$. We write, instead of Equation 80, $\mathbf{z}_{(l)} \in \mathcal{K}^{(l)}\left([\mathbf{A}]^H \widetilde{\mathbf{p}}, [\mathbf{A}]^H [\mathbf{A}]\right)$. There are a variety of iterative methods that, for each iteration l, produce vectors $\mathbf{z}_{(l)}$ that belong to the Krylov subspace l (at each iteration $\mathbf{z}_{(l)}$ can be written as Equation 80). For that reason these methods are called *Krylov Subspace methods* (see Hanke (1995)). It is not obvious, but it can be shown that in general the Krylov subspace methods will approximate the least squares solution $\mathbf{z}_{LS}$ to Equation 79 as l increases (see Hanke (1995), Valdivia & Williams (2005)). CGNE is the most representative Krylov subspace method for the solution to the normal equations in Equation 79. There is also a wide variety of well known Krylov subspace methods like Landweber or Generalized Minimal Residual (GMRES) (see Hanke (1995) for more information). When the matrix $[\mathbf{A}]$ in Equation 76 is square (the dimensions $M = N$) then Krylov subspace methods like GMRES are more commonly used. These methods are based directly in the solution to Equation 76 where $\mathbf{z}_{(l)} \in \mathcal{K}^{(l)}(\widetilde{\mathbf{p}}, [\mathbf{A}])$ (see Calvetti et al. (2001a;b; 2002)).
Despite the simplicity of the CGNE algorithm, each iterate $\mathbf{z}_{(l)}$ satisfies

$$\| [\mathbf{A}] \mathbf{z}_{(l)} - \widetilde{\mathbf{p}} \|_2 \leq \| [\mathbf{A}] \mathbf{h} - \widetilde{\mathbf{p}} \|_2, \tag{81}$$

for all vectors $\mathbf{h} \in \mathcal{K}^{(l)}\left([\mathbf{A}]^H \widetilde{\mathbf{p}}, [\mathbf{A}]^H [\mathbf{A}]\right)$. Another important property of the CGNE algorithm is that

$$\| [\mathbf{A}] \mathbf{z}_{(l)} - \widetilde{\mathbf{p}} \|_2 \leq \| [\mathbf{A}] \mathbf{z}_{(l-1)} - \widetilde{\mathbf{p}} \|_2, \quad \|\mathbf{z}_{(l-1)}\|_2 \leq \|\mathbf{z}_{(l)}\|_2. \tag{82}$$

The proof of these properties is rather technical, so we recommend references Engl et al. (1996); Hanke (1995) for more details.

5.3 Regularization parameter choice methods

We utilize the following definition for our presentation of parameter choice

$$\mathbf{z} = [\mathbf{A}]^{\dagger} \mathbf{p}, \quad \mathbf{z}_{\alpha} = [\mathbf{A}]^{\#} \widetilde{\mathbf{p}}$$

where $[\mathbf{A}]^{\dagger} = \left([\mathbf{A}]^H [\mathbf{A}]\right)^{-1} [\mathbf{A}]^H$ is the generalized inverse and $[\mathbf{A}]^{\#}$ is the operator of the regularization method.

The most widespread parameter choice method is the discrepancy principle, usually attributed to Morozov. This method is based on a good estimation of the error $\|\mathbf{e}\| = \delta_e$ so we solve the problem

$$\| [\mathbf{A}] \, \mathbf{z}_\alpha - \widetilde{\mathbf{p}} \|_2 = \delta_e, \tag{83}$$

for a parameter α.
There are very successful methods that do not depend on the estimation of noise. Two of the most successful methods are Generalized Cross-Validation (GCV) and the L-curve analysis. In GCV we define the function

$$\mathcal{G}(\alpha) = \frac{\| [\mathbf{A}] \, \mathbf{z}_\alpha - \widetilde{\mathbf{p}} \|_2^2}{\operatorname{trace}\left([\mathbf{I}] - [\mathbf{A}] \, [\mathbf{A}]^{\#} \right)^2}, \tag{84}$$

and the optimal regularization parameter α is the parameter that is the minimum of the function $\mathcal{G}$.
In the L-curve analysis we define the parametric curve $(x(\alpha), y(\alpha)$ where

$$x(\alpha) = \log \| [\mathbf{A}] \, \mathbf{z}_\alpha - \widetilde{\mathbf{p}} \|_2, \quad y(\alpha) = \log \| \mathbf{z}_\alpha \|_2,$$

and the optimal parameter will be the point α where the curvature

$$\kappa(\alpha) = \frac{x' y'' - x'' y'}{((x')^2 + (y')^2)^{3/2}} \tag{85}$$

attains its maximum. Here the differentiation $'$ is with respect to α.

5.3.1 Direct regularization

For direct regularization we utilize the following simplifications that come from the SVD, to obtain

$$\| [\mathbf{A}] \, \mathbf{z}_\alpha - \widetilde{\mathbf{p}} \|_2^2 = \sum_{i=1}^{N^*} (f_i^{j,\alpha} - 1)^2 |\mathbf{u}_i^H \widetilde{\mathbf{p}}|^2,$$

$$\operatorname{trace}\left([\mathbf{I}] - [\mathbf{A}] \, [\mathbf{A}]^{\#} \right) = \overline{M} - \sum_{i=1}^{N^*} f_i^{j,\alpha},$$

$$\| \mathbf{z}_\alpha \|_2^2 = \sum_{i=1}^{N^*} f^{j,\alpha} \frac{|\mathbf{u}_i^H \widetilde{\mathbf{p}}|^2}{\sigma_i^2},$$

where the filter factors $f_i^{j,\alpha}$ describe the direct regularization methods defined previously. The GCV and L-curve can be applied directly using the above formulae.
For the Morozov's discrepancy principle in acoustics it was proposed that the error $\|\mathbf{e}\|_2$ be estimated in the following manner. We can define the SNR dB level of tolerance T of our measurements, that under laboratory controlled conditions we can expect $T = -40$ dB. Define the decreasing function $\gamma_i = 20 \log_{10}(\sigma_i / \sigma_1)$ and we denote as i' the index where $\gamma_i < T$, for $i > i'$. Then σ_0 can be approximated by the mean value of the coefficients $|\mathbf{u}_i^H \widetilde{\mathbf{p}}|$ for $i > i'$. This gives the estimate

$$\|\mathbf{e}\|_2 \approx \frac{\sqrt{\overline{M}}}{N^* - i' + 1} \sum_{i=i'}^{N^*} |\mathbf{u}_i^H \widetilde{\mathbf{p}}|. \tag{86}$$

5.3.2 Iterative regularization

For iterative methods like CGLS and LSQR if there is a good approximation of δ_e, then in the discrepancy principle, the method will iterate while

$$\| [\mathbf{A}] \mathbf{z}_{(l)} - \widetilde{\mathbf{p}} \|_2 \leq \delta_e. \tag{87}$$

To apply GCV (as suggested in Hansen (1998)), we use the approximation to the function $\mathcal{G}$

$$\mathcal{G}(l) = \frac{\| [\mathbf{A}] \mathbf{z}_{(l)} - \widetilde{\mathbf{p}} \|_2^2}{(M-l)^2}. \tag{88}$$

For L-curve Hansen & O'Leary (1993), we plot the coordinates

$$x(l) = \log\left(\| [\mathbf{A}] \mathbf{z}_{(l)} - \widetilde{\mathbf{p}} \|_2 \right), \quad y(l) = \log\left(\| \vec{z}_{(l)} \|_2 \right), \tag{89}$$

which gives a curve that resembles an L-shape. The optimal iteration l_{opt} is the iteration with coordinates $(x(l_{opt}), y(l_{opt}))$ which are closer to the point of maximum curvature of the L-shaped curve.

Finally we discuss the rule designed by Hanke and Raus Hanke & Raus (1996). In the CGNE algorithm (Table 3) or LSQR algorithm (Table 4), for iteration l we include the sequence

$$\gamma_{(l)} = \left(\frac{\alpha_{(l)}\beta_{(l-1)}}{\alpha_{(l-1)}} + 1 \right) \gamma_{(l-1)} - \left(\frac{\alpha_{(l)}\beta_{(l-1)}}{\alpha_{(l-1)}} \right) \gamma_{(l-2)} + \alpha_{(l)}, \tag{90}$$

where $\gamma_{(0)} = \gamma_{(-1)} = 0$. The optimal iteration l_{opt} is the iteration for which the function

$$\mathcal{H}(l) = |\gamma_{(l)}|^{1/2} \| \vec{r}_{(l)} \|_2, \tag{91}$$

has its minimum value.

6. Numerical examples

In this section we use numerically generated data to help explain some of the numerical difficulties involved in the NAH technique. Here Γ, Γ_0 and the source surface Γ_s are spherical surfaces. Let the acoustic constants in air be denoted as $c = 343m/s$ and $\rho = 1.21kg/m^3$. The wave-length given in units of m is defined as $\lambda := c/f$.

We consider the problem of recovering the acoustic field on the surface Γ from pressure measurements on Γ_0. The acoustic field is generated at a point $\vec{x} \in \mathbb{R}^3$ for a dipole using the formula in Williams (1999)

$$\begin{cases} p(\vec{x}) = & i\rho\omega Q \left\{ \vec{\alpha} \cdot \nabla\Phi(\vec{x}, \vec{z}) \right\} \\ v(\vec{x}) = & Q \frac{\partial}{\partial \vec{n}(\vec{x})} \left\{ \vec{\alpha} \cdot \nabla\Phi(\vec{x}, \vec{z}) \right\} \end{cases} \tag{92}$$

where Q is the source strength with m^3/s units, $\vec{z}$ is the source location and unit direction $\vec{\alpha} = (\alpha_1, \alpha_2, \alpha_3)$. We denote as $\mathbf{p}$ the complex column vector of M entries, where each entry corresponds to a point $p(\vec{x}_i)$ where $\vec{x}_i \in \Gamma_0$. Similarly we denote as $\mathbf{p}^s$, $\mathbf{v}^s$ the complex column vectors of N entries where each entry corresponds to a point $p(\vec{x}_j)$, $v(\vec{x}_j)$, $\vec{x}_j \in \Gamma$. We compute $\widetilde{\mathbf{p}} = \mathbf{p} + \mathbf{e}$ where

$$\mathbf{e} = \left(10^{-D/20} \| \mathbf{p} \|_2 / \sqrt{M} \right) \mathbf{e}_u,$$

where $\mathbf{e}_u$ is the column vector with Gaussian entries (with normal distribution) and D the signal-to-noise ratio (SNR) used to simulate measurement errors.
The Fourier theory suggests that the maximum distance between adjacent points in the discretization of any of the surfaces Γ_0, Γ, Γ_s for a frequency f needs to be less than $\lambda/4$ to avoid any aliasing problem.
We quantify the numerical error in our reconstructions using the relative errors

$$\|\mathbf{p}^s - \mathbf{p}_r^s\|/\|\mathbf{p}^s\| \times 100\% \quad \text{or} \quad \|\mathbf{v}^s - \mathbf{v}_r^s\|/\|\mathbf{v}^s\| \times 100\%$$

where $\mathbf{p}_r^s$,$\mathbf{v}_r^s$ are the pressure and normal velocity fields reconstructed using our NAH schemes described in previous sections.

6.1 Boundary element methods

For BEM the distance between adjacent points in an element should be less than $\lambda/10$ to avoid aliasing problems and guarantee a proper integration over the element. We consider as Γ the unit sphere with the following discretization points

$$\vec{x_1} = (0,0,1), \quad \vec{x_2} = (1,0,0), \quad \vec{x_3} = (0,1,0)$$
$$\vec{x_4} = (-1,0,0), \quad \vec{x_5} = (0,-1,0), \quad \vec{x_6} = (0,0,-1)$$

and triangular elements

$$\triangle_1 = [\vec{x_1},\vec{x_2},\vec{x_3}], \quad \triangle_2 = [\vec{x_1},\vec{x_3},\vec{x_4}], \quad \triangle_3 = [\vec{x_1},\vec{x_4},\vec{x_5}], \quad \triangle_4 = [\vec{x_1},\vec{x_5},\vec{x_2}]$$
$$\triangle_5 = [\vec{x_6},\vec{x_3},\vec{x_2}], \quad \triangle_6 = [\vec{x_6},\vec{x_4},\vec{x_3}], \quad \triangle_7 = [\vec{x_6},\vec{x_5},\vec{x_4}], \quad \triangle_8 = [\vec{x_6},\vec{x_2},\vec{x_5}].$$

The surface Γ_0 will have the same point distribution as Γ, but Γ_0 will be a sphere of radius 1.1 m.

6.1.1 Higher order elements

At $f = 100$ Hz the wavelength $\lambda = 3.43$ m, and $\lambda/10 = 34.3$ cm. We refine our initial triangular discretization $\triangle_1, ..., \triangle_8$ using the quadratic element scheme shown in Fig. 2a). The first refinement will produce 18 points and 24 linear triangular elements and 6 quadratic triangular elements. In Table 5 we show the relative error for the reconstruction of the pressure $\mathbf{p}$ in Γ_0 and pressure $\mathbf{p}^s$ in Γ that results from a dipole with position $\vec{z} = (0,0,0.7)$ and direction $\vec{\alpha} = (1/\sqrt{3}, 1/\sqrt{3}, 1/\sqrt{3})$ (given in Equation 92), using the direct formulation

$$\mathbf{p}_r = i\rho\omega \left[\boldsymbol{\mathcal{S}}_{(\Gamma_0,\Gamma)}\right] \mathbf{p}^s - \left[\boldsymbol{\mathcal{D}}_{(\Gamma_0,\Gamma)}\right] \mathbf{v}^s, \quad \mathbf{p}_r^s = i\rho\omega \left[\boldsymbol{\mathcal{S}}^{+}_{(\Gamma)}\right] \mathbf{p}^s - \left[\boldsymbol{\mathcal{D}}^{+}_{(\Gamma)}\right] \mathbf{v}^s,$$

for successive refinements. Each refinement reduces the maximum element diameter by half each time. Notice that the recovery of $\mathbf{p}$ utilizes the non-singular matrices $\left[\boldsymbol{\mathcal{S}}_{(\Gamma_0,\Gamma)}\right]$, $\left[\boldsymbol{\mathcal{D}}_{(\Gamma_0,\Gamma)}\right]$ and the recovery of $\mathbf{p}^s$ utilizes the singular matrices $\left[\boldsymbol{\mathcal{S}}^{+}_{(\Gamma)}\right]$ and $\left[\boldsymbol{\mathcal{D}}^{+}_{(\Gamma)}\right]$.
In Table 5 the order refers to the ratio between refinement errors. In the theory found in the book of Atkinson (1997) this ratio is 2^h, where h is the order. The theory also states that linear elements for a smooth surface like a sphere will have $h = 1$, quadratic elements $h = 2$ and cubic elements $h = 3$. The results in Table 5 show that in practice this order can be higher, and this has always been the major motivation for the use of higher order elements for accurate solutions of the forward problem. The use of higher order element will not be a good idea

N	linear Non-singular(order)	linear Singular(order)	quadratic Non-singular(order)	quadratic Singular(order)
18	57.5929	42.7929	39.6083	30.3930
66	17.6444(3.2641)	12.7465(3.3572)	29.2035(1.3563)	22.1742(1.3707)
258	7.5391(2.3404)	5.6551(2.2540)	4.7439(6.1560)	4.5844(4.8368)
1026	2.0327(3.7089)	1.5385(3.6757)	0.2107(22.5112)	0.8449(5.4260)

Table 5. relative error comparison for linear and quadratic elements

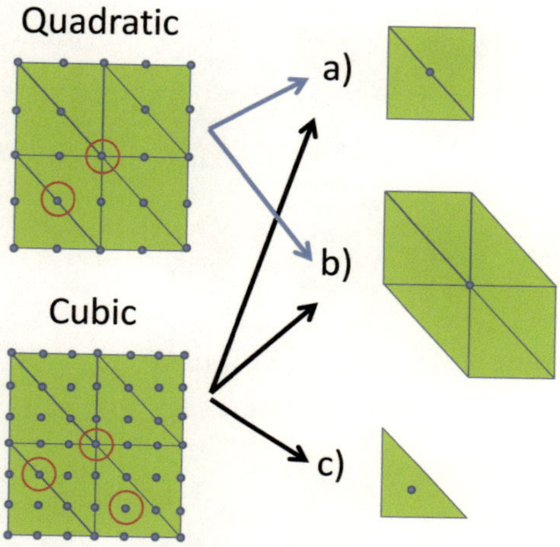

Fig. 6. Triangular decomposition for quadratic and cubic elements. a) multiple triangle intersection for vertex nodes, b) simple edge intersection and c) interior triangle point.

for the inverse problem, where increasing the resolution will pay the penalty of reducing the smoothness of the solution, as we will show.
Let the quantities $W_{m,j}$, $m = 1, ..., n_q$,

$$W_{m,j} = \int_0^1 \int_0^{1-\xi_2} N_m(\xi_1, \xi_2) J(\xi_1, \xi_2) d\xi_1 d\xi_2,$$

for the triangular elements shape functions given in Table 3. These quantities can be thought as approximations of $\mathcal{S}^i_{(m,j)}, \mathcal{D}^i_{(m,j)}, \mathcal{K}^i_{(m,j)}, \mathcal{H}^i_{(m,j)}$ when the field point $\vec{x}_i$ is far from the element $\triangle_j$. We can consider $W_{m,j}$ as an estimate of the column normalized weights (CNW) for the matrices $\left[\boldsymbol{\mathcal{S}}_{(\Gamma_0,\Gamma)}\right]$, $\left[\boldsymbol{\mathcal{D}}_{(\Gamma_0,\Gamma)}\right]$, $\left[\boldsymbol{\mathcal{K}}_{(\Gamma_0,\Gamma)}\right]$ and $\left[\boldsymbol{\mathcal{H}}_{(\Gamma_0,\Gamma)}\right]$.
The following explanation of the CNW will be given for the matrix $\left[\boldsymbol{\mathcal{S}}_{(\Gamma_0,\Gamma)}\right]$, but will also apply to $\left[\boldsymbol{\mathcal{D}}_{(\Gamma_0,\Gamma)}\right]$, $\left[\boldsymbol{\mathcal{K}}_{(\Gamma_0,\Gamma)}\right]$ and $\left[\boldsymbol{\mathcal{H}}_{(\Gamma_0,\Gamma)}\right]$. Notice that $\left[\boldsymbol{\mathcal{S}}_{(\Gamma_0,\Gamma)}\right]$ is obtained by a reduction of the sum $\mathcal{S}^i_{(m,j)}$ and this reduction depends on the triangular elements decomposition used for the surface Γ. Fig. 6 shows quadratic and cubic triangular elements. Observe that the case shown in Fig. 6b) is when a point is intersected by six triangles. This means that when the sum of the quantities $\mathcal{S}^i_{(m,j)}$ are reduced into a column of the matrix $\left[\boldsymbol{\mathcal{S}}_{(\Gamma_0,\Gamma)}\right]$, this point will be associated with the CNW of 6. Using the same argument for the case in Fig. 6a) a column of the matrix $\left[\boldsymbol{\mathcal{S}}_{(\Gamma_0,\Gamma)}\right]$ that corresponds to this point will be associated with the CNW of 2.

Finally, in the case of Fig. 6c), a column of the matrix $\left[\mathcal{S}_{(\Gamma_0,\Gamma)}\right]$ that corresponds to this point will be associated with the CNW of 1. As a result when cubic elements are used, the columns of matrix $\left[\mathcal{S}_{(\Gamma_0,\Gamma)}\right]$ associated with interior points like Fig. 6c will have a CNW of 1. Cubic and quadratic elements for edge points like Fig. 6a) will have a CNW of 2. All elements will have a CNW of 6 in the vertex points.

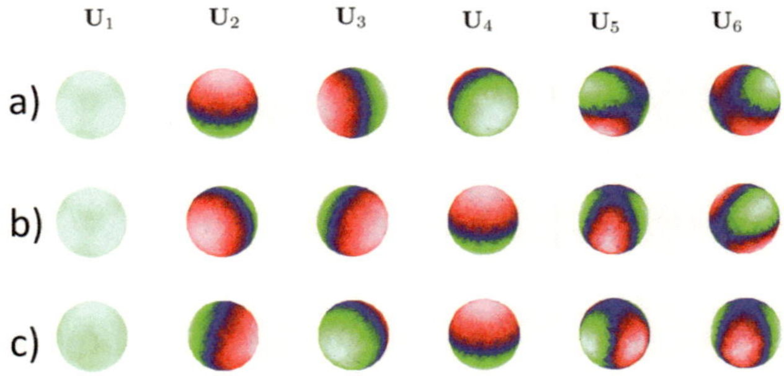

Fig. 7. Plot of the first columns of [**U**] from the SVD of $\left[\mathcal{S}_{(\Gamma_0,\Gamma)}\right]$ at $f = 100$ Hz. a) linear elements, b) quadratic elements, and c) cubic elements.

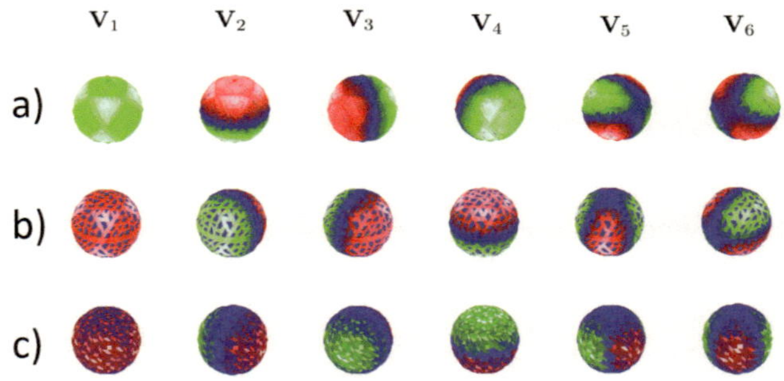

Fig. 8. Plot of the first columns of [**U**] from the SVD of $\left[\mathcal{S}_{(\Gamma_0,\Gamma)}\right]$ at $f = 100$ Hz. a) linear elements, b) quadratic elements, and c) cubic elements.

For the reasons exposed in the paragraph above, linear elements will have a constant CNW while quadratic and cubic elements will have an imbalance CNW. The SVD of $\left[\mathcal{S}_{(\Gamma_0,\Gamma)}\right]$ at $f = 100$ Hz is shows in Fig. 7 and Fig. 8. The imbalance of the CNW for quadratic and cubic elements will produce "zero" dots in the columns of the matrix [**V**] as seen in Fig. 8. Notice that the plots of the columns of [**U**] show that these modes are smooth.

6.1.2 direct regularization methods in action

It was found in the previous section that the use of higher order elements will have a negative influence in the inverse problem. For that reason we will just consider discretizations with linear elements. In this section we will include noise to the measurements and study the effect of noise in the normal velocity reconstructions. Here Γ_0 and Γ are spheres of 1026 points defined as in the previous subsection, and we use the same dipole data at $f = 100$ Hz to

	Transfer Function					Single Source				
SNR	LS	TK	TKH	TSVD	CGLS	LS	TK	TKH	TSVD	CGLS
∞	8.21	8.13	8.18	8.21	8.21	11.94	7.83	11.87	11.94	11.92
40	45.54	11.17	12.29	13.64	12.94	39.92	8.03	12.13	13.80	13.26
30	> 100	15.97	15.45	17.50	16.72	> 100	9.67	13.20	16.23	15.30
20	> 100	26.50	24.10	26.96	25.85	> 100	16.94	18.60	23.94	21.75
10	> 100	42.89	38.36	39.90	40.88	> 100	31.79	30.28	35.18	35.85

Table 6. Relative error for the normal velocity reconstruction using the transfer function and single source matrix systems with different SNR levels. The reconstruction system is solved using LS inversion and different regularization methods like Tikhonov (TK), Tikhonov with High-pass filter (TKH), Truncated SVD (TSVD) and CGLS.

generate the measurements and compare the reconstructions. In Table 6 we show the normal velocity reconstruction relative errors for 2 different systems of the NAH problem, like the transfer function in Equation 53 and the single source in Equations 57-58. For each system we use the conventional LS solution Equation 77 and 4 regularization methods: Tikhonov (TK), Tikhonov with Highpass (TKH), Truncated SVD (TSVD) and CGLS. In each regularization method we utilize the optimal regularization parameter, i.e., the regularization parameter with minimal relative error. Notice that when a lower SNR noise is added, then the errors dominate in the LS solution. For that reason is important to use regularization methods to obtain useful reconstructions.

Finally in tables 7 and 8 solve the same systems than Table 6 using the regularization parameter choice methods discussed in section 5.3: Morozov's discrepancy principle (MDP), Generalized Cross Validation (GCV) and L-curve (LC).

SNR	Tikhonov			Tikhonov with Highpass			TSVD	
	MDP	GCV	LC	MDP	GCV	LC	MDP	GCV
∞	8.5379	8.5379	8.3218	8.2066	10.5212	8.1918	8.2066	11.3525
40	19.6985	19.6985	17.9174	12.4270	13.0708	23.9702	13.7205	13.7475
30	30.5376	30.5376	30.4623	16.0863	17.6388	42.1274	18.5756	18.5133
20	40.6067	40.6067	36.7362	23.9023	25.0874	47.7316	27.0894	27.1740
10	56.3087	56.3087	42.6911	42.9117	39.5407	44.1020	44.7310	41.4242

Table 7. Relative error for the normal velocity reconstruction using the transfer function matrix system with different regularization methods and Morozov's Discrepancy Principle (MDP), Generalized Cross-Validation (GCV) and L-curve (LC) regularization parameter choice rule.

6.2 BEM and ESM

As shown in the previous subsections, the BEM produces accurate reconstructions in noise-less examples. When the measurements are contaminated by noise, then this accuracy is lost and these methods can be compared to the ESM. We utilize the same setup for Γ_0 and Γ as in the previous subsection, and we define Γ_s as a sphere of radius 0.9 m (since the maximum diameter between triangular elements is about 10 cm) with the same point distributions as Γ. As in the previous subsections we use the same dipole data at $f = 100$ Hz to generate the measurements and compare the reconstructions. In Table 9 we show the normal velocity reconstruction relative errors for 2 different methods for the NAH problem, the single source with the modified BEM approach in Equations 64-65, and the the single source with

SNR	Tikhonov			Tikhonov with Highpass			TSVD	
	MDP	GCV	LC	MDP	GCV	LC	MDP	GCV
∞	11.7515	11.7515	11.7593	11.9404	12.4217	11.8831	11.9404	12.8008
40	18.6586	18.6586	16.8357	12.8603	13.8236	22.4443	14.0582	14.0609
30	27.7760	27.7760	26.7282	13.8372	16.4718	38.9839	17.1767	17.4955
20	37.9266	37.9266	31.9003	20.4552	22.4737	45.1214	28.7443	26.0270
10	51.3872	51.3872	32.2178	42.4343	34.2850	42.9328	51.8863	40.0332

Table 8. Relative error for the normal velocity reconstruction using the single source matrix system with different regularization methods and Morozov's Discrepancy Principle (MDP), Generalized Cross-Validation (GCV) and L-curve (LC) regularization parameter choice rule.

ESM in Equations 70-71. For each system we use the conventional LS solution Equation 77 and 4 regularization methods: Tikhonov (TK), Tikhonov with Highpass (TKH), Truncated SVD (TSVD) and CGLS. In each regularization method we utilize the optimal regularization parameter. Notice that we get similar results than the previous section.

	Modified BEM Single Source					ESM Single Source				
SNR	LS	TK	TKH	TSVD	CGLS	LS	TK	TKH	TSVD	CGLS
∞	0.06	0.24	0.18	0.06	0.32	1.28	1.20	1.27	1.28	1.27
40	28.32	4.85	4.65	5.03	4.87	31.86	4.60	4.37	4.91	4.59
30	92.41	9.59	9.36	10.51	9.92	> 100	8.81	8.32	9.57	8.97
20	> 100	20.06	19.29	21.31	20.24	> 100	18.67	17.33	18.89	17.84
10	> 100	36.65	36.15	39.10	38.07	> 100	33.88	32.84	35.71	34.46

Table 9. Relative error for the normal velocity reconstruction using the modified BEM single source system and ESM single source matrix systems with different SNR levels. The reconstruction system is solved using LS inversion and different regularization methods like Tikhonov (TK), Tikhonov with High-pass filter (TKH), Truncated SVD (TSVD) and CGLS.

SNR	Tikhonov			Tikhonov with Highpass			TSVD	
	MDP	GCV	LC	MDP	GCV	LC	MDP	GCV
∞	1.2698	1.2698	1.2698	1.2756	1.2739	1.2754	1.2756	1.2791
40	5.2121	5.2121	6.4921	6.4214	4.4152	8.4317	7.0922	5.0159
30	10.3119	10.3119	10.4154	16.6583	8.9566	12.4129	17.5379	9.6286
20	18.9262	18.9262	18.1611	33.9335	17.2431	18.0954	35.3276	19.7559
10	33.2015	33.2015	38.4855	51.4125	31.7035	33.0507	54.2738	35.1825

Table 10. Relative error for the normal velocity reconstruction using ESM single source matrix system with different regularization methods and Morozov's Discrepancy Principle (MDP), Generalized Cross-Validation (GCV) and L-curve (LC) regularization parameter choice rule.

In Table 10 the reconstruction error is given for the ESM single source matrix system as in Table 9 using the regularization parameter choice methods discussed in section 5.3: Morozov's discrepancy principle (MDP), Generalized Cross Validation (GCV) and L-curve (LC). Finally Fig. 9 shows the semi-convergence phenomena discussed in section 5.2, by plotting the normal velocity reconstruction relative error vs iteration of the CGLS method for the modified BEM single source system. Notice that we plot the optimal iteration and the iteration found by the Hanke-Raus method for different SNR.

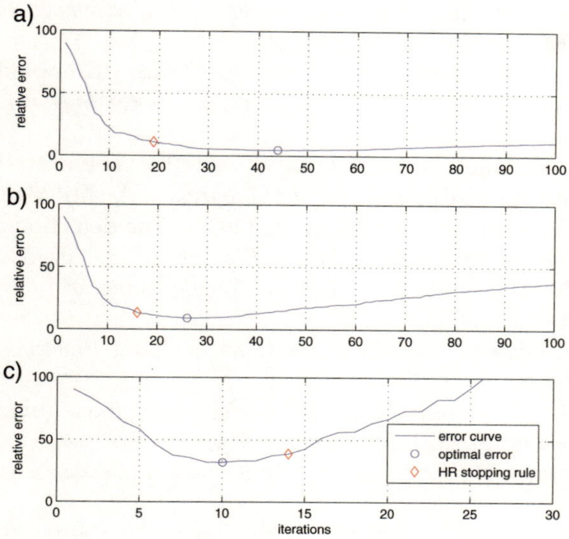

Fig. 9. Normal velocity reconstruction error vs CGLS iteration. a) 40 dB SNR, b) 30 dB SNR, and c) 10 dB SNR.

7. Acknowledgment

This work was supported by the Office of Naval Research.

8. References

Atkinson, K. E. (1997). *The Numerical Solution of Integral Equations of the Second Kind*, Cambridge University Press, New York, NY.

Augusztinovicz, F. (1999). Application and extension of acoustic holography techniques for tire noise investigations, *J. Acoust. Soc. Am.* 105 A: 1373.

Bai, M. R. (1992). Application of bem (boundary element method)-based acoustic holography to radiation analysis of sound sources with arbitrarily shaped geometries, *J. Acoust. Soc. Am.* 92: 533–549.

Borgiotti, G., Sarkissian, A., Williams, E. G. & Schuetz, L. (1990). Generalized nearfield acoustic holography for axisymmetric geometries, *J. Acoust. Soc. Am.* 88: 199–209.

Calvetti, D., Lewis, B. & Reichel, L. (2001a). GMRES, L-curves, and discrete ill-posed problems, *BIT* 42(1): 44–65.

Calvetti, D., Lewis, B. & Reichel, L. (2001b). On the choice of subspace for iterative methods for linear discrete ill-posed problems, *Int.J.Appl.Math.Comput.Sci.* 11(5): 1069–1092.

Calvetti, D., Lewis, B. & Reichel, L. (2002). On the regularizing properties of GMRES, *Numer. Math.* 91(4): 605–625.

Chien, D. (1995). Numerical evaluation of surface integrals in three dimensions, *Mathematics of Computation* 64(210): 727–743.

Colton, D. & Kress, R. (1983). *Integral Equation Methods in Scattering Theory*, Wiley-Interscience Publication, New York, NY.

Colton, D. & Kress, R. (1992). *Inverse Acoustics and Electromagnetic Scattering Theory*, Springer-Verlag, New York, NY, p. pp. 19.

Delillo, T. K., Isakov, V., Valdivia, N. & Wang, L. (2000). Computational methods for the detection of the source of acoustical noise, *Proceedings of ASME2000*, Orlando, Florida.

DeLillo, T. K., Isakov, V., Valdivia, N. & Wang, L. (2001). The detection of the source of acoustical noise in two dimensions, *SIAM Journal of Applied Math.* 61(6): 2104–2121.

DeLillo, T. K., Isakov, V., Valdivia, N. & Wang, L. (2003). The detection of surface vibrations from interior acoustical pressure, *Inverse Problems.* 19(3): 507–524.

Engl, H. W., Hanke, M. & Neubauer, A. (1996). *Regularization of Inverse Problems*, Kluwer Academic Publishers, Boston.

Hanke, M. (1995). *Conjugate Gradient Methods for Ill-Posed Problems*, Kluwer Academic Publishers, Boston.

Hanke, M. & Raus, T. (1996). A general heuristic for choosing the regularization parameter in ill-posed problems, *SIAM J. Sci. Comput.* 17(4): 956–972.

Hansen, P. C. (1998). *Rank-Deficient and Discrete Ill-Posed Problems*, Siam, Philadelphia, PA, chapter 5-6.

Hansen, P. & O'Leary, D. (1993). The use of the l-curve in the regularization of discrete ill-posed problems, *SIAM J. Sci. Comput.* 14(6): 341–373.

Kang, S.-C. & Ih, J.-G. (2000a). On the accuracy of nearfield pressure predicted by the acoustic boundary element method, *J. Sound Vib.* 233: 353–358.

Kang, S.-C. & Ih, J.-G. (2000b). The use of partially measured source data in near-field acoustical holography based on the bem, *J. Acoust. Soc. Am.* 107: 2472–2479.

Kang, S.-C. & Ih, J.-G. (2001). Use of nonsingular boundary integral formulation for reducing errors due to near-field measurements in the boundary element method based near-field acoustic holography, *J. Acoust. Soc. Am.* 109: 1320–1328.

Kim, B.-K. & Ih, J.-G. (1996). On the reconstruction of the vibro-acoustic field over the surface enclosing an interior space using the boundary element method, *J. Acoust. Soc. Am.* 100: 3003–3016.

Kim, B.-K. & Ih, J.-G. (2000). Design of an optimal wave-vector filter for enhancing the resolution of reconstructed source field by nah, *J. Acoust. Soc. Am.* 107: 3289–3297.

Kim, G.-T. & Lee, B.-H. (1990). 3-d sound source reconstruction and field reprediction using the helmholtz integral equation, *J. Sound Vib.* 136: 245–261.

Langrenne, C. & Garcia, A. (1999). Integral formulations for the vibroacoustic characterization of a cello, *J. Acoust. Soc. Am.* 105 A: 1088.

Maynard, J. D. (1988). Acoustic holography for wideband, odd-shaped noise sources, *Proceedings Inter-noise '88*, Avignon, France, pp. 223–231 (Volume I).

Maynard, J. D., Williams, E. G. & Lee, Y. (1985). Nearfield acoustic holography (nah), i. theory of generalized holography and the development of nah, *J. Acoust. Soc. Am.* 78: 1395–1413.

McLean, W. (2000). *Strongly Elliptic Systems and Boundary Integral Equations*, Cambridge University Press, New York.

Nelson, P. A. & Yoon, S. H. (2000). Estimation of acoustic source strength by inverse methods: Part i, conditioning of the inverse problem, *J. Sound Vib.* 233: 643–668.

Ouellet, D., Guyader, J. L. & Nicolas, J. (1991). Sound field in a rectangular cavity in the presence of a thin, flexible obstacle by integral equation method, *J. Acoust. Soc. Am.* 89: 2131–2139.

Pan, J. & Bies, D. A. (1990). The effect of fluid-structural coupling on sound waves in a enclosure - theoretical part, *J. Acoust. Soc. Am.* 87: 691–707.

Raveendra, S. T., Vlahopoulos, N. & Glaves, A. (1998). An indirect boundary element formulation for multi-valued impedance simulation in structural acoustics, *App. Math. Modelling* 22: 379–393.

Sarkissian, A. (1990). Near-field acoustical holography for an axisymmetric geometry: a new formulation, *J. Acoust. Soc. Am.* 88: 961–966.

Sarkissian, A., Gaumond, C. F., Williams, E. G. & Houston, B. H. (1993). Reconstruction of the acoustic field over a limited surface area on a vibrating cylinder, *J. Acoust. Soc. Am.* 93: 48–54.

Schuhmacher, A., Hald, J., Rasmussen, K. B. & Hansen, P. C. (2003). Sound source reconstruction using inverse boundary element calculations, *J. Acoust. Soc. Am.* 113(1): 114–126.

Schwab, C. & Wendland, W. L. (1992). On numerical cubatures of singular surface integrals in boundary element methods, *Numerische Mathematik* 62(3): 343–369.

Seybert, A. F., Soenarko, B., Rizzo, F. J. & Shippy, D. J. (1985). An advanced computational method for radiation and scattering of acoustic waves in three dimensions, *J. Acoust. Soc. Am.* 77: 362–368.

Strout, A. (1971). *Approximate Calculation of Multiple Integrals*, Prentice-Hall, New Jersey.

Sureshkumar, S. & Raveendra, S. T. (2001). An analysis of regularization errors in generalized nearfield acoustical holography, *2001 SAE noise and vibration conference*, number 2001-01-1616, Grand Traverse, MI.

Tekatlian, A., Filippi, P. & Habault, D. (1996). Determination of vibration characteristics of noise sources solving an inverse radiation problem, *Acustica* 82: 91–101.

Valdivia, N. & Williams, E. G. (2005). Krylov subspace iterative methods for boundary element method based near-field acoustic holography, *J. Acoust. Soc. Am.* 117(2).

Valdivia, N. & Williams, E. G. (2006). Study of the comparison of the methods of equivalent sources and boundary element methods for near-field acoustic holography, *J. Acoust. Soc. Am.* 120(6).

Veronesi, W. A. & Maynard, J. D. (1989). Digital holographic reconstruction of sources with arbitrarily shaped surfaces, *J. Acoust. Soc. Am.* 85: 588–598.

Vlahopoulos, N. & Raveerdra, S. T. (1998). Formulation, implementation and validation of multiple connection and free edge constraints in a indriect boundary element formulation, *J. Sound Vib.* 201: 137–152.

Williams, E. G. (1997). On green functions for a cylindrical cavity, *J. Acoust. Soc. Am.* 102: 3300–3307.

Williams, E. G. (1999). *Fourier Acoustics: Sound Radiation and Nearfield Acoustical Holography*, Academic Press, London, UK, chapter 3.

Williams, E. G. (2001). Regularization methods for near-field acoustical holography, *J. Acoust. Soc. Am.* 110: 1976–1988.

Williams, E. G., Houston, B. H., Herdic, P. C., Raveendra, S. T. & Gardner, B. (2000). Interior NAH in flight, *J. Acoust. Soc. Am.* 108: 1451–1463.

Williams, E. G. & Maynard, J. D. (1980). Holographic imaging without the wavelength resolution limit, *Phys. Rev. Lett.* 45: 554–557.

Wu, S. F. & Zhao, X. (2002). Combined helmholtz equation-least squares method for reconstructing acoustic radiation from arbitrarilty shaped objects, *J. Acoust. Soc. Am.* 112: 179–188.

Yoon, S. H. & Nelson, P. A. (2000). Estimation of acoustic source strength by inverse methods: Part ii, experimental investigation of methods for choosing regularization parameters, *J. Sound Vib.* 233: 669–705.

Zhang, Z., Vlahopoulos, N., Allen, T. & Zhang, K. Y. (2001). A source reconstruction process based on a indirect variational boundary element formulation, *Engineering Analysis with Boundary Elements* 25: 93–114.

Zhang, Z., Vlahopoulos, N., Raveendra, S. T., Allen, T. & Zhang, K. Y. (2000). A computational acoustic field reconstruction process based on a indirect boundary element formulation, *J. Acoust. Soc. Am.* 108: 2167–2178.

For EU product safety concerns:
IN TECH d.o.o., Prolaz Marije Krucifikse Kozulić 3, 51000 Rijeka, Croatia,
info@intechopen.com or visit our website at intechopen.com.